D0209136

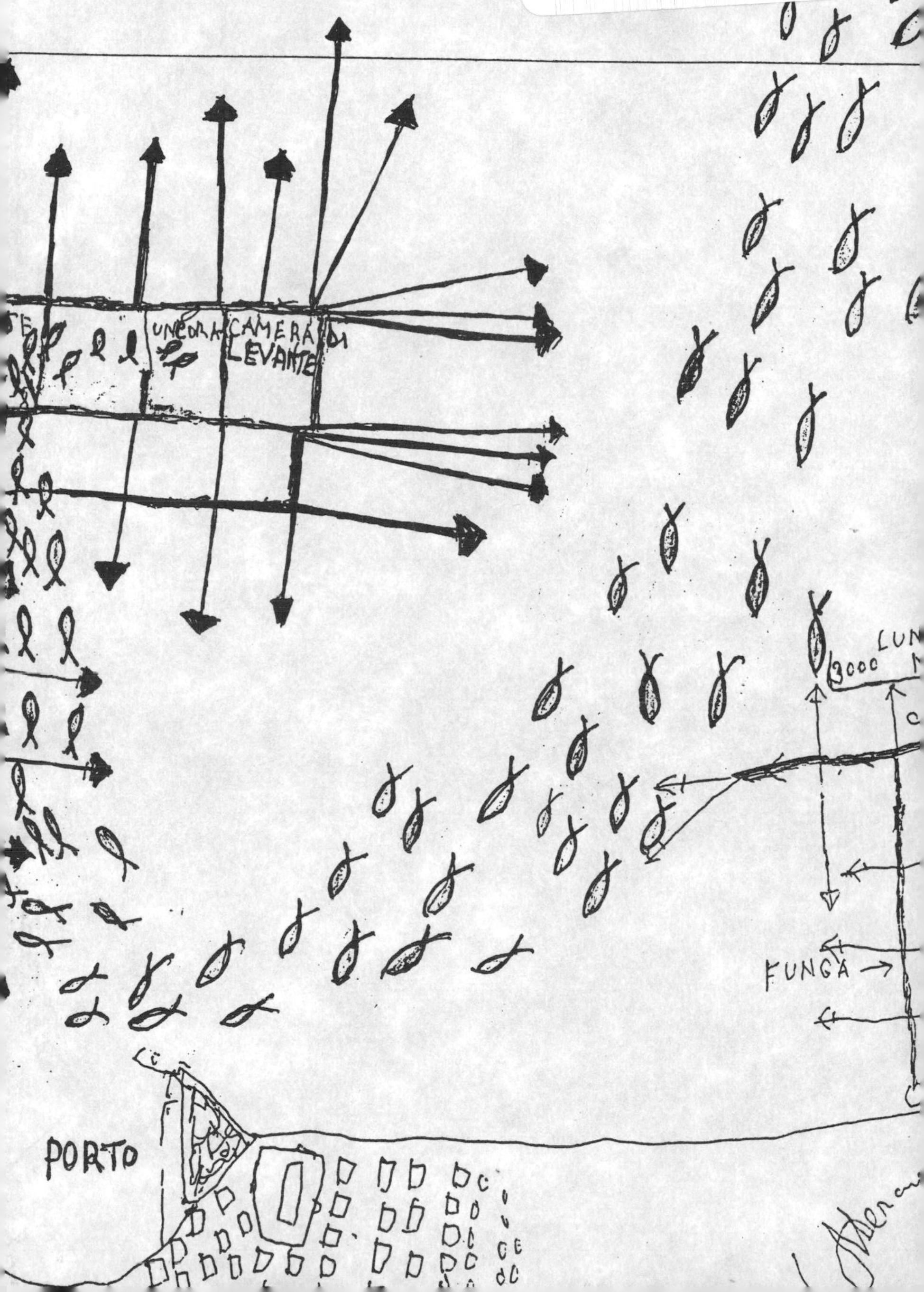
UNCORA
CAMERA DI
LEVANTE
PORTO
FUNGA

Mattanza

Mattanza

Love and Death in the Sea of Sicily

Theresa Maggio

PERSEUS PUBLISHING
Cambridge, Massachusetts

Many of the designations used by manufacturers and sellers to distinguish their products are claimed as trademarks. Where those designations appear in this book and Perseus Publishing was aware of a trademark claim, the designations have been printed in initial capital letters.

A CIP catalog record for this book is available from the Library of Congress.
ISBN 0-7382-0269-X

Perseus Publishing is a member of the Perseus Books Group

Text design by Heather Hutchison
Set in 11.5-point Goudy by the Perseus Books Group

1 2 3 4 5 6 7 8 9 10—0302010099
First printing, February 2000

Find Perseus Publishing on the World Wide Web at
http://www.perseuspublishing.com

To the memory of my mother,
Dorothy Augusta (Botzenmayer) Maggio

Contents

Permission Acknowledgments

Portions of this book were first published in modified form in *Ambassador* magazine, a quarterly publication of the National Italian-American Foundation.

The photograph of the fourth-century B.C. Greco-Sicilian vase "The Tuna Merchant" is reproduced with permission of the Museo Mandralisca, Cefalù, Sicily.

Permission to paraphrase Lucille Craft's article "$20,000 for One Fish?" courtesy of *International Wildlife* magazine, a publication of the National Wildlife Federation.

Chapter 23 photo courtesy of the photographer Antonio Noto.

Permission to quote from Gin Racheli's book *Egadi: Mare e vita* was granted by Ugo Mursia Editore, Milan.

Acknowledgments

Thanks to my father, Joseph A. Maggio, and my Aunt Freda Maggio, for their support, and for not flinching when I quit my day job. Thanks to my sisters, Dorothy Haviland and Marion Zito, and their respective husbands, Pierce and Andrew, for being in my corner from the start. To my late sister Susan and her late husband, Robert: I wish you could be here.

Thanks to my friend, Joyce Marcel, who encouraged me to write this book, for the careful editing. (Yours is next, Joyce.) Thanks to Randy Holhut, for leading the pack; to Jackie and Nicholas Toth, for physical and spiritual sustenance; to Margaret Stearns and to the late Bernard Scholz for letting me live in their homes while I wrote. To Andrea Sweatt Emmons, for Art Night, where it all began.

Thanks to my agent, Sally Brady, who persevered after I'd given up; to Susan Boltz-McCarthy for her wise counsel; to Jody Alessandro, for giving me a chance; to Nancy Newhouse; Erika Goldman; Amanda Cook, my editor, who shaped this book; Mary K. Hartley; Judith Anton; and to Jim Lessing's Family General Store.

In Favignana, thanks to Cristina, Rino, Stefano, and the staff at Bar Due Colonne, my piazza home. To

Rosetta Messina and her family, who opened their home to me; to Antonio Noto; Rosa Manuguerra; Antonio Casablanca; Domenico Messina; and the librarian Giovanna Venza.

In Vermont, thanks to the staffs of the Dummerston branch of the Vermont Department of Libraries and the Brooks Memorial Library in Brattleboro. And to Norman Runnion, my former editor at the Brattleboro Reformer.

In Mondello, thanks to Piero Corrao and Professore Michele LoMonaco for teaching me Italian.

Thanks to Rais Salvatore Spataro and Rais Gioacchino Cataldo for allowing me on the boats and for complete access to the tonnara, and to Rais Gioacchino Ernandes for the interviews.

Thanks to the people of Favignana who shared their stories with me. And my deep gratitude to the tonnaroti of Favignana, past and present.

I have changed some names to protect the privacy of a few people.

Author's Note

The essence of life is that it lives by killing and eating; that's the great mystery that the myths have to deal with," Joseph Campbell said in *The Power of Myth*. In Paleolithic times, "the beasts were seen as envoys from a world beyond the visible," and, Campbell surmised, "a magical, wonderful accord" grew between the hunter and the hunted, as if they were locked in a "mystical, timeless" cycle of death, burial, and resurrection.

A myth opens the door "to the wonder, at once terrible and fascinating, of ourselves and of the universe." "The only mythology that is valid today is the mythology of the planet—and we don't have such a mythology," Campbell said.

Yet I found such a myth still alive on a small island in the middle of the Mediterranean Sea.

Mattanza

Baptism

Piero, who loved me very much, first took me to Favignana. He was thirty-eight, a fisherman of Mondello, near Palermo. He was a good man in a dying trade, a fisherman to the core of his soul. His boat, his nets, and the sea were his life.

I had met Piero a couple of years before when my father and I made a tour of Sicily. Dad is the youngest son of Sicilian immigrants from Santa Margherita Belice, an inland mountain town fictionalized as Donnafugata in

Giuseppe Lampedusa's novel *The Leopard.* We were headed for the ancestral village but decided to stop first for a few days in Mondello, a seaside village, where he had served in the Second World War.

I was thirty-three then, a science writer for the Los Alamos National Scientific Laboratory. Reporting for the lab's house organ, I wrote about physics, materials science, chemistry, the plutonium plant, supercomputers, laser fusion, chaos theory, "Star Wars," and the underground testing of nuclear weapons. Sometimes I wrote morale-boosting stories on the theme of "Bomb designers are people too," accompanied by photographs of scientists in jeans tying flies, bottling home brew, or choosing a bouquet of flowers at the supermarket. I liked the job because it let me watch big science up close, where there were things I never could have seen without my security clearance. Once I watched a buried nuclear bomb explode from the control room fifteen miles from Ground Zero at the Nevada Test Site.

But after a few years a fear of terrorism gripped Los Alamos. I was no longer allowed to humanize a bomb designer by naming his children because they might be kidnapped. About the same time lab security forces installed huge, reinforced-cement anti-tank flower boxes across the entrance to the administration building.

On vacation in Sicily it was easy to forget all that. Dad and I took rooms at Pensione Esplanade, which entitled us to use the private beach reserved for tourists. The strand was a mile and a half long. I had no idea where we were supposed to go. There was a man walking ahead of us; I tapped him on the shoulder. He had his T-shirt wrapped around his neck like a scarf, and my finger dimpled his warm, brown skin. It was Piero. I showed him our beach tickets. He signed for us to follow.

In the summer, when fishing was poor, he was the lifeguard at the tourist beach. He was cocky and strutted like a rooster. He had a big nose, but so do I. His voice boomed like a parade drum. He fell in love with me that day, I don't know why or how. And I thought I loved him too.

I thought I was in love with Piero, but I was in love with Sicily. I went back to see him that fall, and again the next summer. He wanted me to move in with him. Once, in the summer, next to the cornflower sea, I asked him, "But what would we do in the winter?" He said, "We'll stay at home, cook pasta, and steam up the windows." I went back to Los Alamos, gave six months' notice, and numbered the days till I could be in Sicily again.

I read somewhere that women fall in love with the place and marry the man. I never married Piero, but two years after we met I moved into his apartment in a cobblestone street on the north coast of Sicily. He made his living simply, pulling fish from the sea with a net in a small open boat named *Francesca*. There was a mountain at one end of our narrow lane, and at the other end, between the three-story buildings, if you stood on the balcony, you could see a slice of blue sea.

Piero got up in the dark to haul in his net, and I went with him with a pillow under my arm so I could go back to sleep on the sea. The *Francesca* rocked quietly. Hand over hand, he pulled in his net, clotted with algae. He piled the damaged lengths of net in one heap and the lengths holding fish in the other. The pile with tears was always taller. Scavenging dolphins poked holes in the nets, and sometimes they caught on a rock and ripped.

Piero spent most of his time with his nets, making them, mending them, dying them tan so fish wouldn't see them. They were delicate instruments. The cords had to be fine enough and the weave open enough that the current could not carry the net away. They had to be strong yet light enough for him to handle alone. He fished in the big blue horseshoe of Mondello Bay, curled in the arms of beach and mountains.

This is what Piero found in his net every morning: six or seven spiny pink scorpionfish; a couple of sea urchins; fish skeletons; three or four rainbow *viole*; some half-eaten corpses; the occasional lobster; and a kilo of useless sea vermin. His catch half filled one bright blue bucket.

Piero promised me when I moved in with him in February that in May he would take me to see something wonderful, something incredible. Something that made his eyes light up as he talked about it.

We were in our kitchen, and I stood in front of a pot of boiling water. It was a late afternoon in winter, already dark. The windows were black and frosted with steam.

He was gesticulating before me, flapping his arms.

"What is it?" I asked in my elementary Italian.

"Fish!"

What else.

"Big fish." More flapping. "Tuna. Giants." Then he fanned his face with the fingers of both hands many times, very fast. "La mattanza."

This is what I gathered after talking in sign language, Sicilian dialect, my totally useless French, and the Italian I had taught myself from a high school textbook back in Los Alamos: Piero wanted to take me to Favignana, an island off the other side of Sicily, to watch men kill giant tuna in May. There would be much splashing. I'd get wet, all wet.

Piero paid for my Italian lessons with a sack of mackerel. I learned quickly and after just three months was writing for the local monthly newspaper. But as soon as I could speak his language Piero and I began to fight. All he could think of or talk about was fish. But I loved Sicily, so I stayed.

In May, without a telephone call, through some ethereal link between fishermen, Piero learned that the Favignana tuna trap had been set. I never believed Piero would leave Mondello and his nets, but we packed two small bags and left the next morning. We took a city bus into Palermo, then a coach two hours west to Tràpani, spent a night in a pensione, and boarded the first ferry the next morning. At seven-thirty we landed at Favignana and took a room. Piero had brought me here to see a killing, and that was all I knew.

We waited days. Half a century before, Piero's father had killed tuna in Mondello, so he knew what was going on. "They are waiting for their trap to fill," he said.

So on the second day of June 1986 I found myself standing in a seventy-five-foot open wooden boat with no motor. "It never had a motor," Piero said. The boat was black and shiny with pitch, and we were being towed to sea. A couple hundred people were crowded on its deck. The sun was yellow and hot, the sea smooth as oil. A castle on a mountaintop seemed to watch us. The island's cliffs slid by slowly as we glided out of port.

After ten minutes we were a mile and a half out to sea. The boats formed three sides of a square on the perimeter of the trap somewhere in the water below us. Some fishermen lifted one edge of a net and secured it along the gunwale of our boat. The cords that held it up were twisted fiber ropes, frizzled, golden, thick as a man's arm. Palm fronds waved on a pole to the east, three hundred yards distant. We waited for hours.

More fishermen in another longboat like ours a hundred yards east closed the square. The men on it stood silhouetted against the sun, pulling up the other end of the net. As they pulled, the sea inside the square turned from navy blue to lapis lazuli to luminescent turquoise.

After a while huge black shapes rose up into the backlit square. Their slow rising was mystical, like a birth. They rose higher. Dorsal fins swirled, wild animals drawn up from a silent abyss.

They were giants, eight feet long, some bigger, and there were hundreds of them. The net was drawn taut, and they skittered in front of us, half out of the water. I looked into their glassy black eyes. The fish were as big as men, some bigger than four men. When their tails slapped the water it rose in columns above our heads. I remember the din, the thunder of falling water, and their frantic thrashing. They darted to the corners of the net, but there was no way out.

The crowd went wild. People were soaked, screaming and cheering. Piero was delirious with joy. These fishermen were his heroes; their net was full of fighting giant bluefins. It was a scene he saw in his dreams, but he was awake and this was real. Piero tried to pull me back from the edge, but I was riveted. The fish were churning the sea into a white froth, and then the froth turned pink.

When the thrashing calmed they were battered, bleeding and floating on their sides, but they were still alive. The men leaned over the side of their longboat and reached out with barbed gaffs to snag the tuna nearest them. A group of men worked one giant tuna into an upright position against the side of the boat, with the fish's head out of the water. This took ten minutes. Once they gaffed it they held it vertical, half out of the water, and rested a moment.

When the dying tuna quivered it shook the eight men who held it. Its form was a perfect giant pointed egg. Its skin was polished marble, blue as the earth from space. It changed colors as it died. Shimmering veils of opalescent pink and green washed over it. Red blood streamed from a gash in its flank. The men got ready to lift.

They counted to two and heaved the fish to the gunwale, balanced it there on the fulcrum, and rested. One man grabbed a dorsal fin and one a ventral, so when the fish raised its tail again, they used its own impetus to heave it headfirst into the hold, falling forward to avoid the blow. The fish thumped out its life like thunder.

This killing went on for an hour; the blue square turned red. When the last fish was taken, the currents cleared the square of the blood and milky water that clouded it. After five minutes the fishermen lowered the net again into limpid blue water. The dead tuna, in their long black boat, were towed away to another shore, and the tourists, in ours, were towed back to port. No one spoke for a long time. What had I just seen? An eruption, a paroxysm. A font of primal energy, beauty, and suffering, all in a tiny square of sea. It was shocking, and most beautiful.

We packed, caught the ferry to Tràpani, the bus to Palermo, another to Mondello, and went home.

Piero and I eventually split up. But I kept going back to Favignana in the spring. At first I just wanted to see this strange, primitive spectacle, to feel again the strange mix of emotions it stirred in me. Then I had to find out what it took to set up such an enormous trap, and how it came to be, and I went back year after year to learn.

The mattanza was Piero's finest gift, a fisherman's dream, and he gave it to me. He brought me to that slaughter as an act of love, so I saw it as a thing of beauty.

The Cave of the Bluefin

The oldest paintings in Italy lie deep in a cave on Levanzo, a small island three miles off the north coast of Favignana. They are four-thousand-year-old sepia-ink cave paintings of humans and animals: dancing men and limbless, violin-shaped women; equines, bovines, two boars, six fish, and, at the very bottom, the unmistakable diamond shape of a giant bluefin tuna.

Like clockwork every spring, great schools of giant bluefin have appeared in the Gulf of Cadiz off Spain and passed between the Pillars of Hercules to spawn in the Mediterranean. *Thunnus thynnus*, blue-backed, sides of glinting silver, they swim a hundred miles a day in their spring migration. Some come from unknown waters in the eastern Atlantic, some up from the west coast of Africa, a very few from the Americas. They live to be twenty and thirty years old; the largest giants can weigh three-quarters of a ton.

Once they were numberless. The bluefin were to ancient Mediterranean peoples what the buffalo was to the American Plains Indian: a yearly miracle, a reliable source of protein from a giant animal they revered, one that passed in such numbers that the cooperation of an entire tribe was needed to kill them and preserve their meat. Around the Mediterranean the migrating bluefin was a staple food for entire civilizations.

From Spain to Turkey, every spring, men waited for them, first with boats to ambush them and drive them into shallow water, later with harpoons and increasingly sophisticated traps. They watched for dorsal fins from towers built for this purpose. The Greeks had a verb that meant "to watch for tuna." The traps all came to have the same basic elements: barrier nets across the tunas' migration path that funnel them into the mouth of a submerged net cage. In Italy a tuna snare is called a *tonnara*, from *tonno*, the Italian word for tuna.

Harvesting enough bluefin to feed a community for a year required planning and cooperation, so the communities chose one man to supervise the tuna harvest. In Arab countries he was called the *rais* (pronounced RAH-ees), which means "commander" or "head." He was a master fisherman and something of a shaman—one, they all hoped, on good terms with the unseen forces of nature. The title of "rais" became hereditary.

Towns grew up on tonnara sites. The bluefin left its Latin name, *cete*, which means "giant," on ancient maps: Cetaria, now called Scopello, and Terra Cetaria, which extended from Capo San Vito to Segesta, in Sicily; la Cetaria Domitiana, now Porto Santo Stefano on the Italian coast north of Rome, and Cetara of Salerno, site of a modern tonnara; Cetabriga in Spain; Cetabora in Portugal; Ceuta before Gibraltar; Sète in southern France.

The net cage became a many-chambered box where arriving tuna could be sectioned off and corralled until the day of killing. *La mattanza*, the "slaughter" (from the Spanish *matare*, "to kill"), was accompanied by ceremony and song. In all the languages of the Mediterranean people the last room in the trap was the Chamber of Death.

The bluefin seek their birthplace to spawn. It is possible that some of the captured tuna that swim into Favignana's trap began life there when their parents, in a last-ditch effort to procreate, ejected their sperm and eggs as they were being killed. The square of death quickens with clouds of their milt and eggs. Sex, death, and begetting mingle in this briny vessel of primordial juices.

In Italy only a handful of tonnaras still exist. Their numbers have steadily declined, their traditions and rites replaced by the fatal efficiency of long lines and purse seine nets, the modern equivalent of the nineteenth-century frontiersman's repeating rifle. With it a few men could massacre hundreds of buffalo from a train window, transforming the Indians' sacrament into a white man's sacrilege.

Only two tonnaras remain in Sicily, at the foot of Italy. The most beautiful of all is Favignana's, in the Egadi archipelago, one of Levanzo's two sister islands. You can see Favignana's mountaintop castle from the mouth of the cave on Levanzo. The bluefin have always passed between these two islands. Favignana's abundance and stark beauty, and the care lavished on the tonnara buildings—palaces really—made it the Queen of the Tonnaras, the spiritual center of tonnaras everywhere.

At Favignana the tuna fishermen, the *tonnaroti*, still build their trap in the same time-honored sequence every spring. They observe the rituals, recite the prayers, and chant the songs. As the trap slowly takes shape, the tonnaroti become galvanized by the erotic charge set up by unseen thousands of tuna swimming toward them to mate.

The tonnara's central rite, the mattanza, has become part of the life cycle of both hunter and hunted. It is a bloody act of tragic beauty essentially unchanged since the Stone Age.

This way of life cannot last much longer. At Favignana the tuna are smaller and fewer every year. The *ciurma*, the group of fishermen commanded by the rais, has been reduced from a hundred men to eighty, and now to sixty-three, because of mechanization and the need to economize. The bluefin and the tuna traps both will disappear someday. But for now the trap is still set in April, and the wheel of life, death, and rebirth still spins every spring on that tiny island.

For a few years I arrived with the tuna in the spring.

Ai-a-mola

After my year with Piero, I moved back to the States and got a job at a paper in Vermont. Twice I took my vacation in the spring to photograph the mattanza. Then I quit my day job to travel and write.

In May 1993 I went back to Favignana for ten days. I lived in a white cement igloo at the Camping Egad and rented a bike from Zu Isidoro in Via Mazzini, just off the piazza. *Zu* means "uncle" in Sicilian, a term of respect and endearment for the elderly.

The first person to befriend me was Caterina, a twelve-year-old girl from Marsala whose father was the groundskeeper at the campground. When Caterina came to visit him on weekends we biked around the island together.

Favignana is shaped like a butterfly, just nine kilometers from wing to wing, with the port and town at its slim center. The summer landscape is rocky, scorched, and barren, but in the winter, when it rains, the island sprouts a thin green fuzz. Favignana is flat except for a thousand-foot mountain, Santa Caterina, towering above the island's center. The mountain's three peaks bend southward like windblown waves. A castle looms on the highest crest. The mountain's rock runs from ash gray to burnt pink, bare but for a few dark green pine groves. Its north and south cliffs fall in thick folds to the sea. The water sparkles like blue diamonds. Behind the mountain grow prickly pear plantations. On the east wing Caterina and I would stop to pick capers from the bushes that grow wild on the rock outcroppings and rough stone walls.

Caterina shadowed me as much as she could. I was glad for her company because I knew no one on the island. She would sit at my kitchen table and write in my notebook all the words to "Vafanculo," a popular song on the radio. Her mother sent her over once with two strange lemons from their garden. *Limoni lunarie*, moon lemons, crescent-shaped and covered with warty bumps. Caterina dictated her mother's recipe for *limone conzato* for me: "Get a plate. Cut the lemons and squeeze the juice into the plate. Add olive oil, salt (*poco*), and some water until the plate is full. Then lay sliced bread in the plate, let it sop up the juice, and eat."

Life in Sicily seemed like a crystallized dream to me. The sunlight dripped like honey, and the blue of the sea was balm for my soul. Once I told Caterina a dream I had and asked her whether she ever dreamed. “Oh, yes,” she said. “My bed is full of dreams.”

Caterina’s mother would not let her go with me to watch the fishermen work. Every morning at seven the sixty-three tonnaroti gathered outside the Camparia, a group of sandstone block buildings that held the nets, floats, ropes, boats, and cables that made up the trap. The word *camparia* comes from the Sicilian verb *campare*, “to live,” and until the tourists discovered the island, the Camparia had been Favignana’s lifeblood.

They arrived for work on bicycles, on motor scooters, in three-wheeled trucks, and in beat-up cars. They clumped in groups, the older men in the sun, the younger ones in the shade of the Camparia’s high walls, waiting for the gate to open. They smoked cigarettes and talked, gesticulated, and joked. They eyed me, but nobody spoke to me. I was too shy to speak first.

One morning I sat down on a wooden crate outside their gate and flipped through my notebook, trying to screw up the courage to speak to the rais. He was now forty-two years old. When he had taken command six years before, the year I saw my first mattanza, he was the youngest rais anyone could remember. I had a short speech written out. I was looking for it when a man with icy blue eyes in a sleeveless undershirt walked up and offered me a slender green leaf. “Crumple it and smell it,” he said. I did. It was citronella and smelled like lemons. All around me grisly men were sniffing the leaves.

“What are you doing here?” Ice Eyes wanted to know.

“I’ve come to ask the rais permission to follow the men around with my notebook and camera. What is the rais’s name?”

"His name is Salvatore Spataro, like mine, but you just call him Rais.

"We are first cousins," he explained, firstborn sons both named after their paternal grandfather. "They call me Occhiuzzi," which means "small eyes," a nickname he said he'd inherited. Occhiuzzi walked me through the iron grille gate into the courtyard.

The tails and fins of two sharks were nailed to the stone wall of the net house, the shriveled leather remains of two female great whites that had followed the tuna into the trap. Sharks' teeth dangled from gold chains on hairy chests all around. Sun filled the packed-dirt courtyard; its twenty-foot walls of creamy white sandstone were wavy in spots, worn by the wind and rain. Colored, plastic floats and coils of thick rope were scattered on the ground. Directly ahead of me was the net house with its great door wide open, pitch black inside.

Occhiuzzi took me by the arm and walked me over to another man who ushered me politely into the rais's office just inside the gate to the left. The room was dark and smelled of damp stone and mildew. The rais sat in the gloom behind a scarred wooden desk.

I was scared; he was wary. He tapped a pencil on his desk and looked me in the eye. I saw his short legs stretched out under the desk. He slouched in his chair, affecting ease but exuding tension like an engine idling too high. His stubble beard showed salt and pepper, his thick black hair was turning iron gray. Being in charge had aged him.

I stood before him. With a wave of his hand he invited me to sit. My eyes adjusted to the dark. I took a deep breath and recited the speech I'd memorized.

"I'm a journalist. I saw my first mattanza in 1986 when a fisherman from Mondello brought me here. Since then I've seen two more mattanzas, and I've read some books. I know the tonnara is more than one day of killing. I want to learn about it firsthand. I'd like your permission to follow your men around while they work and take notes and pictures."

He smiled at my nervousness. Here I was, a woman alone, in her late thirties, with a beat-up black knapsack on her back. It was my office in a bag, bulking with my Pentax, my down vest, two lenses, pens, film, and an Italian schoolchild's exercise notebook. He could have been annoyed, but he chose to be amused. He was king here, and I was a creature from another planet, a woman alone who did as she pleased. I had dropped from the sky into his world.

"You are weeks too late to see the trap set," he said. "It is already in the water. Now we wait for it to fill." I was crestfallen.

"Be here at one-thirty this afternoon," he said. "Show her around now," he told the man who brought me in. We walked through the empty net house, past a carpenter's shop, through an alcove filled with old black iron floats, into another, hidden courtyard that opened on to the sea. To my left was a huge, round-hulled wooden boat. Riddled with cracks, its white paint peeling, it leaned drunkenly in its dry dock. When the men looked up from their work I answered their glances with a nod.

At noon I walked to the dock, pulled my brown bag of olives, bread, and cheese from my knapsack, and ate at the water's edge. That afternoon I went to sea with the rais to check the trap.

I arrived early and waited outside the Camparia gate for the rais to emerge. I stood alone with knots of fishermen all about me, waiting to go back to work. One man walked toward me, a giant. On my spring trips I had taken many pictures of him as he killed the biggest tuna. He looked to be in his fifties, stood six-foot-five, had a black beard and mustache, broad chest, and a friendly gap in his crooked white teeth. He was bursting with some kind of animal life; the air crackled around him. He bent over and asked me, confidentially, "What do people in America think of the mattanza?"

"Those I tell about it think it is barbaric," I said. He straightened up to his full height. "My name is Gioacchino Cataldo," he said, and waited while I wrote it down. "You tell them this is what is barbaric: to raise a calf, feed him, and fatten him, make him think you're his friend, then one day to slaughter him. The fish, the tuna, he comes to me and I take him. But I never knew the tuna before then. The beast of the land cries, but the tuna, no. There's a butcher here in Favignana. On Wednesdays and Thursdays I change the route I take to work because these are the days he slaughters animals, and passing by there makes my heart tremble. With the tuna, no."

The rais emerged. Without a word, six men fell in behind him and walked to the end of a floating pier where three navy blue boats and the rais's small gray boat awaited. The blue boats had the names of planets and fish painted in white letters on their hulls. One was called *Levanzo*, and one *Favignana*. I couldn't find a name on the rais's gray boat. "It has no name," he said. "This is the *musciara*." He steadied me as I stepped into it.

Other men manned the blue boats. The crew of the musciara took up thick unpainted oars, their paddles only as wide as a man's hand, and rowed us out into open water. Their hands were thickly callused and scarred. They cast lines from stern to bow, and a small tugboat, the only vessel with a motor, towed us in single file. When we picked up speed the two men in the musciara's prow drew a black tarp behind their heads to shield the rest of us from the sea spray.

We bumped over small waves. I had brought a tape recorder and asked the rais to let me record him saying the prayer I'd heard about. The fishermen removed their caps, and the rais began a litany he yelled into the rushing air. He prayed in his dialect.

"Na sarvirriggina a matri ri diu ri Tràpani!"

A Hail Mary to the Mother of God of Tràpani. Ten seconds of silence, time to say the whole prayer silently. The sea slapped the boat five times. The town slipped away. He shouted again:

"A Hail Mary to the Mother of God of the Rosary!" The sea spray hissed.

"A Hail Mary to the Mother of God of Calvary!"

"A Hail Mary to Saint Theresa!" He skidded through the syllables and pressed hard on the last one, making the saint's name toll like a bell and trail off.

"A Hail Mary to the Madonna of Fatima!" The base of the mountain slipped by.

"An Our Father to the Patron Saint Joseph!"

"An Our Father to Saint Francis di Paola!"

"An Our Father to the Sacred Heart of Jesus!"

"An Our Father to Saint Anthony!"

When we could see the tip of Marettimo he lowered his voice and prayed, "An Our Father to Saint Peter, that he pray the Lord for a good catch."

The fishermen shouted, "May God make it so!"

"Eternal rest, Holy Creator, for our dead. Holy good day," he said to his crew. They put their caps back on, and soon we arrived at the western edge of the trap. The towboat cut its engine so as not to scare the tuna. The men took up their oars again and rowed to their positions above the trap's eastern chambers.

The fishermen say that the bluefin behave like sheep, and that a shepherd designed the first tuna trap. The men tied the boats up to cables and opened gates that connected the trap's chambers. A collapsible net door was in the center of each wall between the submerged rooms. When closed the flap was drawn taut to the surface. When open it dropped and furled on the seabed. The men could draw it closed again with ropes attached to it.

Each morning and afternoon they untied the ropes and let the gate fall open. Their job was to herd the fish toward the *bastardella*, the last room before the Chamber of Death, and to close the gate behind them.

The musciara tied up over the bastardella and hovered above the trap. The men stretched a canvas tarp over two bench seats and removed a board that covered a glass window in the floor of the boat. The rais crawled under the tarp to watch for tuna.

When the fish passed through the gate the men had to haul it up, hand over hand, through a hundred feet of water faster than the fish could panic and turn back. The tuna were timid, wary, and lightning fast, so the arduous process had to be repeated many times.

The six-man crew began to forget I was there and talked about food and women while I made myself small and quiet. After a while the rais let me climb under the tarp. I lay on my belly and felt the ribs of the rocking boat press into mine. With my face against the window, I watched in wonder. The last time I saw such a light I was looking down into the core of a nuclear reactor where Cerenkov radiation made the coolant water glow blue. Suddenly three giant bluefins glided across the neon blue screen, right to left, bathed in this hazy radiance. They were serene black shadows, alien beings with no regard for the world above them. The fish circled the room clockwise in small groups at a depth of fifty feet.

Too soon another fisherman took my place under the tarp. I came out and blinked in the light. It was June, the air was cool, but the sun made us drowsy. Cotton-ball clouds scudded above. Some of the men curled up in the curve of the prow and slept, but one always manned the window to count tuna.

"What do the tuna eat?" I asked the rais.

"Triglie," he said. They were mullets, prized fish that Piero loved to find in his net. "No, don't write that." He looked in my notebook to see what I'd just written. "Tell them we're not barbarians," he said. "Tell them we don't have bones in our noses and bones in our hair. We're civilized people."

He said he was just kidding about the mullets. "Tuna eat triglie if they can, but they'll eat anything. It's true, there's not much for them to eat inside the net, but they're not thinking about eating now. They're in love."

The fishermen kept knotted white plastic bags of crusty bread, sharp pecorino cheese, and fruit in a cabinet under the stern; after a while they shared their food with me. They passed around a communal water bottle, tilted their heads back, and let the water pour down their throats without touching their lips. They skinned pears with the same knives they used to cut bait. To pass the time, Vito, the youngest in the rais's musciara, brought a short rod to fish for smallfry. He baited a hook with sardines and lowered it into the water; the man under the tarp watched it through the window and told him when a fish was about to bite.

The ferry passed a few hundred yards from the trap. "Occhio," said the man at the rudder. Watch out. After a minute the musciara lurched and swayed on the ferry's wake. The handsome white ship was a regular nuisance; though distant its noisy engine and churning propeller scared the fish and made them dive; then they became uncountable.

I asked Vito, "How do you know you haven't counted the same fish twice?"

"I can tell one individual from another," he said. To me, from a height of fifty feet, the tuna looked like thin black diamonds, and the hours of contemplating them seemed more a quiet meditation than an accounting.

The men told me their names. The one called Marino drew a wind rose in my notebook and told me the names of the currents and winds. The *tramontana* from the north brought storms and the fish, but recently there had been too much fine weather. The rais was asleep, but his face bespoke worry. The sea was overfished, the tuna smaller and scarcer than just ten years before.

Still, the boats floated above the trap for two hours twice a day, and most afternoons I was with them.

On the seventh day, when the rais was under the tarp, Vito tapped my shoulder, lifted his shirt, and showed me his stab wound. He put his finger to his lips and said he'd been in a fight with a colleague. He tenderly peeled off the white bandage taped above his left hip; the stitches were fresh. He begged me not to tell the rais or he might be removed from the musciara for misconduct. The men knew not to talk, but the rais would find out anyway.

In 1986, at the close of his first season as rais, the headlines in the *Giornale di Sicilia* had read, "Miraculous Catch: 2,551 Tuna Taken at Favignana." Since then, just enough tuna had been caught every year to allow the tonnara to break even. The men of Favignana had set this trap for a thousand years, and Salvatore Spataro would probably be in charge when the end came.

"Too many people picking the fruit," he said.

He had been only thirty-six when he commanded his first mattanza, newly promoted by Franco Castiglione, an entrepreneur from Tràpani who held the lease on the tonnara. The former rais had quit because Castiglione dared to interfere with his command. The men mocked Castiglione; he had too much money and not enough sea sense. One year, they said, he had crashed his small, fast yacht on the Favignana rocks when he mistook a streetlight for a buoy and gunned for it full speed. The next week, "without complaining once," he bought a new yacht. But Castiglione kept their tonnara alive.

Salvatore Spataro was a strong leader from the start. When he had taken command many of the net walls had been made of vegetable fiber. The fledgling rais told Castiglione he needed new nylon nets that would not rot or break so easily, no small investment, and got them in time for his second season. The century-old black wooden boats were in bad shape; instead of fixing them, Castiglione had a new fleet built of iron and painted them navy blue. The old floats that pulled the net walls taut to the surface had been heavy iron balloons in 1986; now the rais had new ones in lightweight plastic, orange and red.

On the ninth morning, just before seven, I bumped over the cobblestones on my bike, on my way to the Camparia. A wooden door to one of the houses in Via Roma was propped wide open. Inside the bare room a woman in black sat on a wooden chair next to a single bed. In the bed lay a white-haired woman, fully dressed, with her eyes closed. She was very still. When I passed again a few hours later she was in a coffin surrounded by women sitting on wooden chairs. Tall flowers leaned in vases set on the tile floor. Men in suits smoked outside at the curb.

When the rais came back to his office from his morning visit to the trap he told his men to be ready for a mattanza the next morning. My money was running out, and I would have to leave right after the first mattanza. But that afternoon they took Girolamo, the tonnara's diver, to check the trap for swordfish, sharks, and holes in the net, and I went with them.

In the musciara I sat facing Benito Ventrone, a fifty-two-year-old tonnaroto who was filling in for Vito. "His grandmother died before dawn," he said.

Benito was a character. He pulled his yellow-white hair back into a thick ponytail and almost always wore a blue bandanna bordered with leaping dolphins tied over his bald pate. He cultivated a snow-white beard and kept it trim and clean. His shirtsleeves had a razor crease from wrist to shoulder and his collar gaped open showing the pound of gold on his neck, rows of thick chains with heavy pendants.

A small boat on a wide sea becomes an intimate place. I leaned over, lifted the gold charms from his neck, and weighed them in my hand: a crucifix three inches tall; a two-inch-wide heart with an alighting dove; a long squiggly charm like a swimming sperm; the face of Christ crowned with thorns; and four gold-mounted ivory triangles with serrated edges—the shark teeth. One gold chain was too long, so he tied it in knots.

We moored at the bastardella, the penultimate chamber. Girolamo sat on the gunwale, strapped a six-inch knife to his right thigh, pulled his face mask over his steel-wool beard, and toppled backward into the sea. The rais followed the bubbles with his eyes.

Salvatore Spataro looked haggard. He was wearing the same faded jeans and the same pink and black checked flannel shirt he'd worn for the past three days. He scratched his beard; he hadn't shaved. To rest his back he knelt in the bottom of the boat and put his head on his arms.

"I could sleep two hours," he said. "I haven't slept for two nights, and I won't sleep tonight either." He hung his arm over the side of the boat, and a thick gold bracelet, his only ornament, dropped down around his wrist. The back of his hand touched the smooth water and caressed it like a woman's cheek. He stared into the water, into the web of light there, then drew back into the boat.

He lay down on the boat floor, his right arm outstretched, palm up, fingers curled, the water still dripping from his knuckles. He was handsome and virile, and no one questioned his command. This was his world, the shades of blue, the castle, the sun broken up in the water, the men who did his bidding.

The men in the next boat were fifty yards away. They shouted, "Sono entrati!" Some fish had passed through the gate, and four men stood up and leaned back, straining on the ropes, taut as tent pegs, swinging their arms in full circles, left arm then right, to pull the gate up in time. The men in the musciara shouted encouragement. The bluefin passed into the bastardella and stayed put.

The rais sat up and asked for the two pink rocks he had brought with him to drop into the bastardella. "The fish will rush to them," he said. He was tired of hearing me call them tuna. "The creatures below us are called fish as long as they are in the water," he said. "Only when they are dead are they called tuna." He tossed the rocks over the side and watched them sink. No fish appeared, but Girolamo surfaced. He bobbed by the side of the boat, lifted his mask, and asked for a rope. Then he dove again and tied the cord to the crescent-shaped tail of a dead bluefin entangled in the net. They hauled it up and lifted it aboard, rubbery and compact, its hide smooth, its curvilinear form perfect. A hundred kilos, they guessed.

At four o'clock we headed back to port. I went to Via Roma to buy a bottle of Marsala wine and some flowers to leave in the rais's office with a note of thanks. I thought I wouldn't have time to do so before catching the ferry the next day. But while I was at the florist, a black minivan came for Vito's grandmother next door. A funeral procession formed to follow the hearse to church. The mourners clutched inverted bouquets of gladioli. I paid for my flowers, but the old florist wouldn't let me leave. Quietly he clicked the door closed, then locked it and lowered the blind. We stood in the dark.

I asked him why.

"This is our custom," he said. After the corpse passed he opened the door.

That afternoon the tonnaroti gathered in their courtyard. The old ones were silent, the younger ones jabbered more than usual. Franco Castiglione rang the bell that announced the year's first mattanza. The rais was out of sight. Occhiuzzi shouted, "Forever be praised the name of Ge-SÙ!" and the men removed their hats and yelled, "Gesù!" three times.

Shortly after dawn the next day I was back at the Camparia. A crowd was waiting for permission to board the longboat. The tonnaroti picked up their gaffs, strong oaken rods tipped with sharp steel barbs. Benito shouldered his ten-foot boat hook and draped one arm over it, like an elephant's trunk draped over a tusk. The fish would become tuna this morning.

I boarded along with a couple hundred of the curious. Promptly at seven we were towed in a long line of boats to the trap. The rais was in his musciara, the last in line. He looked five years younger this morning. He had shaved, got a trim, and changed his shirt. He wore a blue sweatshirt that said "Boy America."

The longboat with tourists sidled up along the western edge of the Chamber of Death. Fishermen assigned to it winched up that side of the net with a windlass and secured it aboard. Imagine a box a hundred yards long and seventy-five feet wide, with no top but with a net inside it at one end, suspended from the rims like a square hammock. That square area is the Chamber of Death. The net hangs a hundred feet deep at its center.

Three smaller boats lined up on the north side of the Chamber of Death, and three boats along the south. The rectangle was now open to the east. I saw men in small rowboats lower the gate between the bastardella and the Chamber of Death. Then they watched and waited for the fish to pass through.

The wait was long. A thin, well-coiffed lady stood next to me chain-smoking Marlboros and throwing the butts into the Chamber of Death. They floated, but she berated a boy who threw his orange peel in the water. Bored and hot, she removed her cotton sweater and sunned in her blue silk tank top. She reached into her purse, drew out her cell phone, and dialed a number. She spoke to a friend and never mentioned that she was three kilometers out to sea about to witness the mattanza. She had no idea what she was in for.

When the fish had passed into the Chamber of Death a towboat pulled a waiting longboat identical to ours into position to close the east end of the quadrangle. The rais and his boatmaster were alone in the musciara in the center of the square. He had his back to us and faced the fishermen and the sun behind them, raised his arms together slowly, and shouted "Aisa!" They started to pull and sing at the same time.

He directed their pulling so the net came up evenly. Forty men stood along the side of their longboat. Each man bent double, dropped an arm, hooked his fingers in the net, stood up to pull it aboard, then bent double again, drawing the net up to the slow rhythm of a song.

Occhiuzzi chanted the verses. The men responded in a deep bass chorus: "Ai-a-mola, ai-a-mola." Pull, you Moor. Pull.

The song was old, and its words were strange. It was an alien tune never meant for a mortal audience. The tonnaroti sang to a god they threatened and implored. The tourists hushed their chatter. The rais shouted to his men, blowing his whistle. The men shouted to each other between choruses. We strained to hear the words blown across the water.

Jesus Christ and all the saints
And the Holy Savior
Who created the moon and the sun
Created all the people,
Created the fish in the sea
The tuna and the tonnaras
A promise is a debt!
This God must help us
And send us into salvation
Calm sea and wind at our backs
To find a safe harbor
To let us find a place out of the storm.
This God must help us
Avoid every evil.
The great birth-giving saint
The Holy Virgin gave birth
And made a son like God
Ai-a-mola. Ai-a-mola.

As they sang, the square got smaller and the fishermen came closer to the crowd. Now the shadowy forms appeared, crowded, circling the perimeter of the chamber, fish still in their element.

A son like God was born
Who was called Jesus.

Jesus, bring me good fortune
One or the other lasts little
One or the other of short duration.
Queen crowned
Queen of this world
Let the dawn bring a beautiful day
A day like we have had before
Like we have received,
Like Easter, like Christmas, which are important feasts,
To Saint Joseph you were wife
You, Saint Joseph, were husband of Mary
And Mary helps us
Because we are her children.
Old Saint Joseph
Carries the ax and the chisel
And in his arms beautiful Jesus.
He carried blessed Jesus.

When they reached a seam where the weave tightened, Occhiuzzi yelled, "Cambia!" then changed songs and quickened the rhythm. He sang "Gnazù," which asks God to bless the rais and his musciara, and cursed the "Turkish dogs" who once plundered tuna and kidnapped tonnaroti in these waters. No one I asked knew what *gnazù* meant; the men sang it after every verse like a Buddhist chant:

Good soldier of the wars (Gnazù!)
Leader from the wars (Gnazù!)
Good-looking men, onward with love! (Gnazù!)

They are roses and flowers (Gnazù!)
They are bright-colored clothes (Gnazù!)
They are precious vestments (Gnazù!)
And may this year be saved. (Gnazù!)

The sea surface bulged, and then dorsal fins cut crazy lines in the satiny water. The fishermen sang "Lina, Lina," a bawdy song about the daughter of the boss. They praised her breasts, her vagina, her lips, her legs, and her hips. She was naked because the catch was poor; her father needed fish to provide her dowry. The singers threatened to give her to the rais.

The small square of sea foamed and boiled, and the men stopped their chant. The net was taut, the water rose up in a wall of sound. The crowd gave a throaty roar. The woman whose silk blouse was now wet and sticking to her shrieked, half delighted, half afraid.

"Spara a tunina!" the rais yelled as the fish appeared. When the foam turned pink, he gave the order to kill.

Gioacchino and his partner gaffed a fish that was bigger than both of them. The men worked in five teams of eight, eight men to each bluefin, four on either side of it. The middle pair had the shortest gaffs, arm-length, and hooked the fish behind the gills. The poles graduated in length, so the men on the outside with the ten-foot gaffs hooked the fish just above the tail. They lifted the fish upright, and when it was standing their poles formed a chevron, pointed toward the sea.

"Uno. *E Due!*" They leaned back, eyes squeezed shut, their faces contorted in grins, their biceps bulging, and heaved the fish to the gunwale. The tuna was theirs now. It balanced a moment on the gunwale, then it raised its tail and lowered it, and the men fell away on top of one another to avoid its blow. The great fish drummed a hollow tattoo on the boat planks.

"Oh, this is cruel," said the woman with the phone. "I won't ever eat fish again. I didn't know it was going to be like this." But like the rest of us, she was transfixed.

The men vied for the biggest bluefins. The exhausted fish swam on their sides toward our boat, their saucer-sized eyes turned up to our faces. They went to the corners to escape but met the drawn-up net; they hid under the musciara but were pulled out with gaffs. I could smell their pungent oil, see their colors run out. It went on for nearly an hour.

When the last tuna was taken, Occhiuzzi shouted, "Forever be praised the name of Gesù!" and the fishermen swung their caps and shouted "Gesù!" in unison. A burly man with shoulder-length blond curls somersaulted into the Chamber of Death. I guessed it was his own private fertility rite. Then others dove into the murky red water and washed the warm blood off their limbs.

They took 161 tuna that day. "Pochi, ma belli," the rais said. Few but handsome. The fishermen threw buckets of crushed ice over the fish that half-filled their longboat. The *Nino*, Castiglione's noisy diesel-engine, towed the tuna to Tràpani while a little tug towed the tourists to land.

I packed up, cleaned my campground igloo, paid the bill, returned my bicycle, and went back to the Camparia in the afternoon to say good-bye to the rais and his crew. The rais shook my hand. "When you return you'll find friends here," he said.

"It was an honor to accompany you in your work," I said.

"The pleasure was mine," he said. "You're not sad to leave?"

"I did my crying in private," I said. I would miss this island. It was a microcosm of Sicily, so strange and beautiful and cruel.

I took the one-thirty ferry to Tràpani and stood on the aft deck to watch Favignana fade away. A mile and a half from shore we passed the red and orange floats that outlined the trap. Five boats were out on the afternoon check, tied up to trap cables, rising and dipping on a small sea. The *Nino* was just heading back to port. I stood on a deck bench and waved both arms and shouted, but they didn't look up. They were watching for giant fish. The ship cut a wide arc around the Chamber of Death, and after a while the fishermen rocked in its wake.

September

I could not bear to see the island recede without promising myself I'd come back soon. In September I arranged a trip to visit my cousin Nella in Santa Margherita Belice, my ancestral village, and to travel around Sicily a little. But first I'd stop for a few days at Favignana.

I stayed at the campground again, in another cement igloo. I had hoped that by mid-September the summer people would be gone and I'd have the island to myself.

But the weather was fine and the tourists were still there, so I extended my stay by two days at a time, determined to sit at my table outside the Two Columns Bar until the last pair of Bermuda shorts and sandals with knee socks was gone and the crowd had sifted down to just locals in flip-flops.

I bought the *Giornale di Sicilia* every morning from Antonio Noto's newsstand and bookstore in the piazza. It was the town's living room—a sunny rectangle bordered by four bars, a few stores, the poultry butcher, the baker, the library, a pharmacist, three-story apartment buildings, and three private men's clubs that comprised a room, a party affiliation, a mess of newspapers on a table, and some ashtrays. In the middle of the square the nut vendor sold roasted garbanzo beans and salted pumpkin seeds from a wheeled cart. A bouquet of helium balloons strained at long strings from its push bar.

The piazza's north end was taken up by the facade of the church, set a few yards left of center to give the Spanish cannon at the old fort up the road a clear shot to the sea. The Spaniards, who once ruled the island, had built the fortress not far from the west end of the piazza, on the remains of an eleventh-century Arab stronghold. The fort protected the tuna trap with artillery pointed at pirate ships come to plunder the tonnara.

When Antonio's father died, he'd taken over the family newsstand, but he was a photographer at heart. All the mattanza posters and the best postcards of the island were his work. One morning Antonio told me that one of the tonnaroti, Clemente, the burly man with the golden curls, conducted private tours of the island's archaeological zones. Clemente was not hard to find. The Two Columns was his bar too, and besides, he stood out in a crowd.

At forty-eight he was massive: thick-boned and built low to the ground. His face was chiseled, his jaw square. His mouth was a thin straight line, his eyes were hooded, and his irises were bright blue. He was the only Clemente on the island. But what set him apart more than anything else were the thick yellow curls that fell to his shoulders and framed his rugged face. This strange combination of masculine and feminine fascinated newcomers, including me.

I'd seen him stand stock-still at the bar with his thick fingers curled around a stem of *prosecco* while strangers stroked his biceps and fingered his hair, as if he were a wax statue. "Yes, it's real. No, I don't use peroxide." He was always patient and polite.

I found him at the Two Columns, having lunch. We agreed to meet at the statue of Ignazio Florio in Piazza Europa at six that evening. Then I went to the beach at Lido Burrone for the afternoon. The sun was hot, but the breeze was cool. Soft waves pushed coral pebbles into a scalloped pink border along the shore.

At six I was waiting when Clemente pulled up astride a huge cream-colored Moto Guzzi motorcycle that purred and growled. He was stonefaced, all business, dressed in loose drawstring pants and a white T-shirt that stretched tight across his chest and middle-age paunch. Without a word of greeting, he nodded for me to get on. He braced his legs against the ground and offered me a cannonball bicep as a handle up. When he took off there was nothing to hold on to but his waist. Long curly tendrils of hair streamed back and tickled my nose. I hunkered down out of the wind behind him; it was like squatting behind a boulder.

The first stop was a small sea grotto on the north side of the island facing Levanzo. The Romans once raised moray eels for their tables here, he said. "For fun, they would put a slave in here and cut his leg so the eels would smell blood and attack. The basin used to be covered with mosaics." In the dark of the cave he spied a couple of dull black marble tiles, small, just a half-inch square. He picked up the tessere and gave them to me. Then he pointed with his chin to a fly-covered pile of human excrement in the back of the cave. "Now it's a bathroom," he said, disgusted.

Tufa outcroppings jutted into the sea there. Tufa is the island's sandstone. In the six centuries before Christ, when the Mediterranean sea level was two feet lower, the ancients cut the coastal tufa for building blocks. The cutting left flat square shelves that were now washed by the sea. Children were wading in the shallow pools on the rocks. Here and there stood sandstone columns too soft to use for building; locals had carved them into giant human figures.

In 241 B.C. the Romans took the Mediterranean Sea from Carthage here, Clemente said. Until then, the Mediterranean had been a Carthaginian lake.

The Carthaginians had inherited their Phoenician ancestors' shipbuilding skills and sailed galleys curved like silver skates and tiered with rows of oars. The Romans had scant experience in naval battles or shipbuilding, so they captured the ship of Hannibal the Rhodian (not the later Hannibal of elephant fame) and made two hundred copies of his *quinquereme*, a war galley with five men on each oar. With this new fleet, the Romans won the Battle of the Egadi Islands. Legend says that Cala Rossa, a small bay on Favignana, was so named because the sea turned red with blood during the battle. The Romans took the ten-mile-wide strait between Favignana and Marsala, ending the First Punic War and setting the course for Western civilization. Favignana had seen the start of the Roman Empire.

Next, Clemente took me to a small, round cultivated field surrounded by semi-submerged caves.

"This is San Nicola. Here it was a city. Understand?" My Italian was better than he thought. We parked and crossed the plowed field on foot, stepping over the concentric furrows. The farmer had unearthed bits of broken amphorae—pink handles and smooth shards that fit the curve of my palm.

"How is it that a field full of amphorae can be plowed up and planted and not preserved for archaeological study?" I wondered aloud.

"No one cares," Clemente said.

Romans lived here after the battle, he said, but it was a Phoenician village in the sixth century before Christ.

"They all lived underground. This was their home, their graveyard, everything. They would dance in this piazza and be protected by the rocks from being seen at sea. No building materials were used, which is why their homes still exist."

Outside each cave entrance someone had drilled a hole in the rock to tie up a horse or cow with a rope or leather thong. The beast's manger was a depression chipped out of the rock. "I have found a lot of bronze fishing hooks in these caves," Clemente said.

I followed him across the field again. He stopped once to scratch the dirt with his toe and picked up a fragment of a Phoenician glass paste vial. I have since seen vials like it in the museum on the nearby island of Mozia, once a Phoenician stronghold. They held the ashes of firstborn children the Phoenicians burned as a sacrifice to their god Ba'al. The shard shone in Clemente's palm, satiny and iridescent. He handed it to me, and I put it in my pocket.

When we came to a crumbling stone wall he scrambled to the other side, turned around, squeezed my shoulders, and lifted me over it. We walked to an ancient well whose opening was no longer a round hole but a six-petaled flower. "The well was owned by six families," Clemente explained, "and each owner dropped her bucket from her own side of the well."

He took me to a low embankment under which was a hole in the ground. He squatted on his haunches and dropped a few feet into the cave. I followed and landed on my feet in a carved room fifteen feet square, edged by a dry canal two feet deep. A crude bas-relief crucifix carved in the wall surmounted a globe and a mysterious death mask. Clemente had no explanation for any of it.

Then he pushed me like baggage back up into the sunlight, climbed out behind me, and bent down to pick up a thin greenish wafer half the size of a dime.

"A Roman coin," he said, and this time stuffed the find into his own pocket. "You're good luck," he said. "Usually I don't even find Phoenician glass."

Clemente said that a hundred years ago, when men first quarried tufa for export, they found Roman coins in the Cavallo shaft but threw them away, thinking they were false and had been planted there to make fools of them.

Back on the Moto Guzzi we followed the coast with the foamy blue sea to our left, around to Cala Torretta to see the Phoenician oven tombs. They looked like small bread ovens carved into the cliff face. Once they held clay jars containing the remains of the dead. In one of these tombs Clemente had found petrified skulls. "Once I found a child's skull with the baby teeth intact," Clemente said. "I left it where it lay."

Back on the motorcycle to visit another set of caves: One had a Spanish coat of arms and a cross carved deeply on one wall; another cave had an inscription in Latin, too worn for me to read; another had a carved inscription written in neo-Carthaginian from the first or second century before Christ. The script looked like a cross between Greek and Arabic. The graffiti were as mysterious to me as Clemente himself. He was tight-lipped and sardonic, but polite. He muttered a minimalist answer for every one of my questions.

He wore a gold chain with a tiny gold Aladdin's lamp.

"What's that pendant on your neck?" I asked.

"An Arab ointment vial. I found it when I was diving."

"Where were you born?"

"In Caserta, near Naples," where his mother was from, but he grew up on Favignana, his father's island.

Nearly two hours had gone by. He spoke in a monotone voice and so reluctantly I thought he wished this tour were over, so I asked no more questions. He went to his bike, toed the kickstand, swung his leg over, and I got on behind him. By eight o'clock the eastern sky was black; the sun had set but was still sending up its colors in the west. Behind the mountain the sky was a parfait of rose, purple, and midnight blue with the castle silhouetted against it. Clemente took me on a sunset tour of the eastern wing of the island. The Moto Guzzi hummed. He leaned back against me and turned his head so I could hear: "This is a fine signora," he said, patting his gas tank.

Back in Piazza Europa I paid him the thirty-five dollars, thanked him, and shook his hand, and he tooled off back to his own life.

I never saw the rais that September. He was a figure of the tonnara and seemed not to exist outside that world. But his cousin, Occhiuzzi, stopped at my table every day as I sipped my morning coffee in the sunny piazza.

Occhiuzzi was married, but I couldn't figure out what he did, if anything, for a living outside of tuna season. He had a boat; probably he fished. He didn't hang out with friends in the piazza, and he was often seen putting off purposefully to places unknown on his little motor scooter.

With no tuna trap to watch, I had lots of free time and spent it biking around the island, looking for idyllic lonely beaches. The beach of the day changed with the winds that swirled around the island. Each morning Occhiuzzi would look up at the weather vane atop the green dome of the *chiesa madre* (mother church) and tell me which beaches would be protected from the wind. But for one reason or another, I never went to the beach he suggested.

He'd find me back in the piazza around four and ask me, "Where were you? I took my boat around, but you weren't there." So I knew he'd been looking for me, and I didn't like it. I wanted only peace, quiet, and sunshine.

The next morning was my last day on that particular trip, so I was in the piazza early. Occhiuzzi parked his scooter and came to my table. This time he gave me directions to a beautiful secret cove, a blue grotto with white pebbles at the foot of the mountain. "You'll have the place all to yourself," he said. I followed his directions, rode my bike as far as it would go behind the cannery, then padlocked it and walked ten minutes along a narrow cliffside path. I found the cove, and it was stunning. I stayed three hours.

With no one around I took off my bathing suit and became part of the tawny landscape. Dwarf palms dotted the hillside. A carved staircase led down to the sea. I sat on the steps and closed my eyes. When I opened them I was in paradise. The water was limpid turquoise. I thought I would burst; I could not contain the beauty. But every once in a while I scanned the horizon for an approaching fishing boat, half expecting Occhiuzzi to invade by sea.

So I was still naked when I heard him yell my name fifty feet behind me. He had come by land, in his bathing trunks, after lunch. I held up my hand like a traffic cop. "*Stop!*" I said. He watched while I put on my bathing suit. Then he sauntered down, towel in hand.

"Nice pony," he said, having seen the horse tattooed on my back. His breath smelled of onions and brandy. I watched his eyes scanning me, sizing up my flaws.

"What are those?" he asked. Four little blue spots in a vertical row on my upper arm. I hadn't even noticed them. I thought for a minute.

"Oh! That was Clemente." Occhiuzzi looked jealous. I explained that Clemente had lifted me over a wall when I hired him to show me the caves. Occhiuzzi's lips grew tight. I had been alone with Clemente. He made his move.

"Look," I said. "I'm a journalist. I want to return here and live here and write a book, and I don't want to have problems."

"A book about Favignana?"

I nodded yes.

"And you'll put everything into it? Even this conversation?"

"Yes."

I invited him to go swimming, to show I wasn't afraid. But I was. A dip in the frigid water would do him good.

"I can't go in the water. I've just eaten," he said.

He dipped his hand in the sea and sprinkled me. He touched my face and smoothed my hair. I froze like a deer in headlights. He backed off.

"It's natural for a man to ask a woman to make love. Right?"

"Right," I said, flatly. No man here would think of denying his animal nature. Making a pass was his duty. A common occurrence here.

"Ti saluto," he said. "I'm leaving, good-bye." But he didn't leave. "You don't want to make love?"

I shook my head no.

"Then I'll go." He offered his hand, and I shook it. I was in an awful position. Occhiuzzi was a *capoguardia*, one of the rais's two lieutenants, as well as the rais's cousin. I needed his goodwill to have access to the tonnara. "You're okay," I said.

"I know. I know I'm a good man." And he went.

A half hour later I biked through town to the port to buy a ticket to Tràpani. I went through the motions like a robot, like the bereaved at a funeral, doing what had to be done. I had found my island, and I wanted to stay forever. "Partire è un pò morire," they say here. Leaving is a little like dying. You leave, and the place goes on living without you.

The campground van brought my bags to the Two Columns, where I had a last beer with Cristina and Stefano, two of the owners, and we said our good-byes. I was crying when I returned my bike to Zu Isidoro. He gave me a discount and said, "You know where to come back to."

I waited until the last minute to run to the port. I arrived all sweaty, bags in hand, and was the last one to board the hydrofoil.

As for Clemente and Occhiuzzi, they did not stir from the piazza, did not even say good-bye, although they both saw me before I left. They knew I'd be back.

The Ropes and the Nets

I worked odd jobs all winter and saved enough money to rent a house on Favignana for forty days in the spring of 1994. At the end of April I went back, this time to see the barrier nets set. As we neared the island I stood on the deck of the *Donatello* and watched for the tuna trap. The island's now familiar silhouette loomed larger, and there, in the dark blue water, were the orange floats, lined up like lights on a runway.

I felt a thrill, waiting in the hold for the ferry's maw to yawn open. At the Two Columns, Cristina greeted me like an old friend. She had seen this happen before: People visit her island once and come back forever after. I know one man who, after his first visit, left his architect's position in continental Italy and moved to Favignana permanently. I sipped my welcome-home Campari and scanned the corkboard covered with postcards from people who wished they were here. Cristina coddled her customers. Like me, many of them became her friends.

I crossed the piazza to the tourist office to see about renting a room in a fisherman's house. "That is not possible," said the woman at the desk. Instead she handed me a typed list of locals who would rent their second homes, an important source of income on Favignana. No prices were listed. It might be too expensive, but I decided to try anyway. I picked a name and dialed the number. Michele Ingrassia said he had just the house for me.

"How much?" I asked.

"You can afford it. Come look at it."

He picked me up in the piazza with his wife and daughter and drove five kilometers to the easternmost point of the island, past old tufa quarries and chocolate-brown horses and a herd of cream-colored cows and fields of poppies and buttercups, clumps of pines, and straight stone walls. We pulled up to an iron gate. Through its bars I saw a stuccoed stone house shining white against a cobalt sea. Across the strait I could see the hazy gray outline of Sicily. The house stood on an island of sandstone in a sea of empty space created by the two exhausted tufa mines that Ingrassia's father and grandfather had quarried.

Michele unlocked the house: a dining room, master bedroom, guest room, indoor bath, outdoor shower, and fully equipped kitchen, with espresso pot, colander, cheese grater, and mortar and pestle for mashing garlic.

A shaded porch fronted the road, and two whitewashed patios opened in the back. A ladder led up to the roof. Fragrant pines, red geraniums, and white calla lilies grew in the well-tended garden. The house came with a small bicycle and tire pump, all mine for only twenty-five dollars a day because it was so far from town. I knew it was worth twice that, but I simply couldn't afford it.

"How much can you afford?" he asked.

"Fifteen dollars a day." And the deal was done. I had to sleep at the campground bunkhouse for one night while Signora Ingrassia cleaned the house. (It looked just fine to me.) They brought me back to the Two Columns so I could relax after my long trip. I saw Angelo, Clemente's twenty-five-year-old son, and asked about his father.

"He's home in bed, not feeling very well," Angelo said. Angelo was tall and thin, with his father's blond curls but not so long. I sent his father my regards. A nun in a gray and black habit entered the cool dark of the bar from a small sunny courtyard out back where a children's party was in progress. She offered me a slice of ice cream cake.

The next morning I was up at six and started walking to the Camparia. Girolamo, the tonnara's diver, already in his wet suit and on his way to work, stopped to pick me up in his open off-road vehicle.

The men had parked their mopeds and three-wheeled trucks against the wall of the boathouse. They had already had their coffee. They smoked Marlboros and stood around in twos and threes, waiting for the gate to open. There was a breeze; seagulls hovered and dipped over whitecaps. On the dot of seven the gateman in his blue felt beret turned a skeleton key in the padlock and swung open the ten-foot iron gate.

I greeted Occhiuzzi. We both acted like nothing had happened, and that was fine with me. I said hello to the men I recognized from the rais's musciara the previous spring. The others shot me furtive glances; they didn't want to stare. I went straight to the rais's dark office, with a new speech.

He sat behind his desk in a flannel shirt and an Elmer Fudd hat with the flaps up, but this time he stood up to shake my hand. Last year, I explained, the tonnara was to be a chapter in a book about little-known places in Sicily. I wrote that chapter, and it came out well. But then I decided the tonnara deserved a book of its own.

"It is such an ancient and unusual event, and you have given me the privilege of seeing it from the inside. I've come for forty days. I want to see the nets go in the water this time, and the anchors, and all that goes before the mattanza. I want to immerse myself in the culture of the tonnara and try to understand it. You may be the last rais. I want to show how the tonnara has changed over the centuries, and with it the character of your job." I'd done some research over the winter.

"Once," I continued, "the authority of the rais was unquestioned, and that, I think, is no longer so. First came the unions, and now you cannot even call a mattanza unless the Japanese freezer ship is on the horizon. Now it is the Japanese who command; the foreign market is in charge."

I hit a nerve. The Japanese bought all the big tuna from Castiglione.

"It is I who command, and you must understand that!" He was incensed.

"But isn't it true that you wait for the arrival of the Japanese freezer ship before you call for a mattanza?"

"I can order a mattanza any time I want," he said.

Now that I had his attention, I decided to press on with other questions I had scribbled in my notebook.

"How do you get along with Castiglione?"

"We get along all right," he said. "Our relationship has been nice and clean."

"Was it always your dream to become rais?"

"This, no. I never thought about becoming rais. It happened. I didn't want to take the responsibility. I was barely thirty-six when I started in 1986."

"Did the men accept your authority? You were so young."

"The men accepted me, even the oldest ones."

"Are there any satisfactions that come with being rais?"

"Enormous satisfactions."

"For example?"

"When I manage to have a beautiful mattanza. When the work goes well."

"What makes a beautiful mattanza?"

"Lots of tuna. In 1989 we took 694 tuna in one mattanza."

In a normal year that would have been more than half the season's catch. He slid open the top right desk drawer and took out a journal in which he had written the years and the number of tuna taken at every mattanza he had commanded.

At that moment Clemente came limping in from the sunshine. He'd let his hair grow past his shoulders. He'd been in an accident; a nun in a hurry had passed him on a curve, and he and his Moto Guzzi went down on some gravel. The huge motorcycle rolled over him twice, maybe his foot was broken. The skin of his upper arms had been scraped off onto the asphalt. The rais told him to take a couple of sick days, and Clemente limped out. Then the rais made a call on his radio phone.

"Should I leave?" I asked.

He grinned. "No. I'm not calling a girl."

I stuck around to watch the work. The oblong skeleton of the trap was already in the water, but not much else was done. Occhiuzzi explained what was happening.

"First we put the skeleton in the water, the floats and the cables, then the anchors and the stones to hold them in place, and then the nets are hung from the cables. It takes two and a half hours to drop two and a half kilometers of net once the skeleton is set up. It's easy."

The trap and the barrier nets had to be broken down every year for repair and storage. Every spring the nets were sewn back together, new stone anchors were cut, and the cables and floats were reattached and carried back out to sea.

The nets were stored opposite the gate in one of my favorite buildings. Its interior looked like the belly of a whale, cavernous, vaulted, and ribbed with high pointed arches, the same shape, strangely enough, as a bluefin's upper jaw. Inside it felt more like a cathedral than a warehouse; tons of mesh mounted up to the thirty-five-foot ceiling.

I wandered in unescorted. Inside the net house it was dusk all day; the light from bare bulbs hanging at intervals from cross beams could not fill the space. The floor was dark red brick. A cloud of soft daylight floated through a clerestory window. The air glittered with motes of dust and dry fish scales.

The hall brightened when the tonnaroti opened the great double door to move a net outside; the twenty wishbone arches caught the glow and seemed to shine. They were the same pointed arches I'd seen in the Palatine Chapel in Palermo and the cathedral of Monreale. I felt like I could levitate under them.

Three men climbed to the top of the western stacks, where the barrier nets were kept, La Coda and La Costa. They threw down a section of rolled-up net, tied at each end by a tight twist of rope. On the floor they stretched it out to its ten-meter length and cut the ropes that bound it. Three of these sections sewn together lengthwise would stretch from sea surface to sea bottom, about a hundred feet. Hundreds of these three-part lengths sewn side by side formed two barrier nets. The Coda was the tail that would extend three kilometers from Favignana's port to the trap's entrance. The Costa would skirt the north coast of the island and extend five kilometers northeast into the strong currents that brought the bluefin to the trap.

The men checked the nets for holes, repaired them, then sewed them together, attaching chains to the bottom, floats to the top. They set out steel cables to hold the net walls up and anchors to hold them in place at sea. These days the nets were made of nylon and the floats were plastic, but once the entire trap was made of sisal and Indian coconut fiber and the floats were Sardinian cork. These required more work—throwing buckets of freshwater on the nets and letting them dry completely before storing them. They rotted fast and had to be replaced. The rais used to have one-third of the trap remade annually. Now the nets were shiny nylon, and only one-fifth of the trap was renewed every year.

I introduced myself to the gateman who took his place inside on a low wood stool next to loops of golden coconut twine. His name was Giuseppe Aiello, and he was sixty-two. He said they still used the coconut rope for a couple of things: tying the stone anchors to the net walls, and for the *ancera*, the curtainlike net they used to scare the tuna into the Chamber of Death. "The coconut net won't tangle," he said. Aiello was working on the anchor ropes. He laid a stiff skein of twine on a stump and chopped each end of the loop with an ax to make equal lengths. I caught a whiff of pine every time he chopped. "Sea pine," he said, and chopped me off a chunk to sniff. I still keep it on my writing table.

He was making thirty-five hundred *capizaghi*, the thin cords that would attach the stone anchors to the nets. Just three years before he had to make seven thousand *capizaghi* for the season, but for the past three years a thick metal chain had replaced almost half of the expensive and irretrievable tufa blocks. Aiello cut the lengths, wrapped an even thinner twine around the frayed end of each cord, and knotted it tightly. Tedious work.

The youngest tonnaroto that year was nineteen years old, the oldest was seventy-four. Some had been fishermen all their lives, some were mariners who had seen the world, others had worked in German mines, some had been factory workers or taken construction jobs in Germany in the winter. Most had had to leave the island to work for a number of years in order to afford a house here. Some now owned several houses and rented them to tourists in August.

The older men, the ones who had been fishermen or sailors all their lives, came to work in berets, blue work smocks, and tattered overalls. The younger men showed up in their grimiest jeans, the same pair maybe three days in a row, or until their mothers or wives were too ashamed to let them leave the house. ("A man's appearance is the mirror of the wife," the women say here.) For work aprons they wrapped burlap bags around their waists and tied them with coconut twine.

One section at a time, they stretched the net the length of the center aisle and cut the twine that kept it folded like an accordion. Four men at each end had tucked serrated table knives in their breast pockets and used them to pick off the knots of rope that held the weights or floats the previous year. It took hours to check the nets for holes. When they found one they yelled "Ferma!" (Stop!) and Occhiuzzi appeared instantly with a fisherman's needle. His blue knit watch cap pushed down over his eyebrows, he set his jaw and darned the hole so fast his hands were a blur.

Occhiuzzi was in a rush, and he was pushing the men. If he had had a whip he would have cracked it. Just that morning they prepared thirteen pieces of net, more than they usually do in an entire day. The tonnaroti, scrambling like ants up and down net hills, burst out in bits of song. They seemed happy and charged with spring. The jokers among them tied each other's shoelaces together and fell over laughing when somebody tripped.

Outside in the bright, windy courtyard two men with their jacket hoods pulled tight around their faces stretched sixty-foot lengths of twine while a noisy gasoline engine twisted them into rope. Every type of cord they used had a name, and so did every thickness of rope, and every tool that twisted, cut, or measured it. Zu Massino coiled a newly made rope and placed it on top of a pile of coils stacked against the courtyard wall. Over the years the ropes had worn a cylindrical groove in the stone wall.

Man and Bluefin

The roots of the tonnara sink into prehistory and spread under the civilizations of the Mediterranean. The four-thousand-year-old cave painting of a giant bluefin on Levanzo is the earliest evidence of tuna hunting in this area. But after the cave paintings the intertwined tale of man and bluefin disappears for thousands of years.

In the fifth century before Christ, Carthaginians and Phoenicians stamped their coins with the bluefin.

Aeschylus's tragedy *The Persians* contains a reference to the tuna hunt. When a messenger recounts to the Greeks the terrible naval defeat of the Persians at Salamina, he says: "The beach and the rocks are full of corpses and the remaining barbarian ships flee in chaos, leaving the Hellenes to beat the ranks of the enemy with bits of oar and flotsam, as they do with captured tuna or with fish caught in a net."

In 350 B.C. Aristotle described the bluefin's voyage from the Atlantic to the coast of Bosporus, where, he affirmed, they reproduced. Pliny the Elder, in the middle of the first century, reported that nurses liked to give the children in their charge meals of ground tuna liver to make them grow strong. Strabo, a Greek geographer of the Roman era, in the first half of the first century said the Phoenicians pushed past Gibraltar to intercept schools of tuna and built a tuna-salting plant at Cadiz in Spain. Plutarch, in the same era, explained that the tuna swam in perfect arithmetical order "to stay together and for love of one another."

In the second century the naturalist poet Oppiano of Cilicia, in his discourse "On Fishing," described a Roman tonnara:

> Dropped in the water are nets arranged like a city. There are rooms and gates and deep tunnels and atria and courtyards. The tuna arrive in great haste, drawn together like a phalanx of men who march in rank: there are the young, the old, the adults. And they swim, innumerable, inside the nets and the movement is stopped only when . . . there is no more room for new arrivals; then the net is pulled up and a rich haul of excellent tuna is made.

The finest tuna were sacrificed to the sea god Neptune. The word *sacrifice* comes from the Latin *sacrum fecere*, "to make holy."

In those times the Mediterranean was an unlimited source of food, for both man and the migrating bluefin. Oppiano wrote of an incredible abundance of sardines, on which the bluefin feed: "The impact with such schools of fish stopped ships as if they'd run up on rocks; at times, despite every force of arms, the oars remained stuck in that mass."

The Byzantines passed the first law protecting a tonnara. It prohibited fishing around private traps, an indication of their economic importance.

In 807 A.D. North African Arabs founded the tonnara of Favignana, although there may have been some undocumented plant there before this date. In the early ninth century new tonnaras were springing up in Sicily, Africa, and Spain. The Arabs gave the tonnaras their music and terminology, but they brought no great technical innovations to the tuna trap.

The arrival of the Normans in Sicily in the eleventh century gave the fishery a big push forward. Ruggero II, the first king of Sicily, claimed the tonnaras for the crown but often leased them to private parties to generate income. The king also instituted tithes for the bishop of Mazara, in the modern province of Tràpani, from all the tonnaras in that diocese. The bishop granted a special dispensation to tonnaroti to work without sin on Sundays during the spring trapping season and threatened excommunication and eternity in hell if his 10 percent was late.

In 1176 William the Good gave the Benedictine monks of Monreale the tonnara on the Isola delle Femmine (the Island of Women), near Palermo. Its proceeds so enriched the abbot that he became the largest landowner in Sicily after the king himself. The salted tuna of Sicily, along with ships' biscuits made from Sicilian wheat, was sailors' fare throughout the Mediterranean. The tonnaras' success kept the windmills turning at Marsala's famous (and still extant) salt pans. In the thirteenth century all the tuna caught at the Sicilian tonnaras of Castellammare, Monte Cofano, Bonagia, San Cusumano, San Giuliano, Scopello, Palazzo, Favignana, San Teodoro, Magazinazzi, Siccara, Carini, Capo Boeo, Mazara, and San Vito was preserved in Marsala's sea salt.

In 1266, under the French Angevin rulers, tonnara production was so high that Sicily began to export salted tuna to Naples. Around 1314 the pirate ships holed up in the hidden coves of the Egadi Islands put the Favignana tonnara out of business for a while. Despite the threat of brigands, setting bluefin traps remained a major Sicilian industry and source of revenue for the king. To keep the money flowing a law was passed declaring that tonnara workers, from the rais to the last shopboy, were immune from legal proceedings, civil or criminal, during the spring bluefin season.

Medieval tonnaroti worked for a wage specified in a seasonal contract, as they do today, and took a percentage of the tuna and all the smaller fish inadvertently trapped.

By 1400, with Sicily now under Spanish rule, bluefin meat was being preserved in olive oil from the island of Jerba, off the coast of Tunis, which in the Middle Ages had sometimes been joined to Sicily politically. Spanish rulers sold royal titles and tonnaras to raise capital. In

1638 the Stella family spent sixty thousand *scudi* acquiring a tonnara near Tràpani and a baronial title. At that time, sixty thousand scudi amounted to far more than what the Spanish king collected from export taxes on Sicily's wheat, its chief crop, in a good year.

Spain, indebted by its war in the Lowlands, borrowed money from Camillo Pallavicino, a Genoan patrician, who accepted the three Egadi Islands, including Favignana and its tonnara, as collateral. In 1637 Pallavicino called in the loan and took possession of the islands.

In the 1800s the heirs of the Marquis Pallavicino married into the Rusconi family of Bologna, and the Egadi Islands became their commonly owned property. In the meantime, at Palermo, Vincenzo Florio was becoming a rising star in the world of finance. Sicily's own Rockefeller, Florio made his first fortune in quinine, the new cure for malaria, and went on to make several more fortunes in disparate industries. He noticed that the wealth of many Spaniards came from the tonnaras, but he also saw that they were run inefficiently, not having been improved technically since Roman times. He believed that with some innovations he could increase profits.

In 1829 Vincenzo Florio bought the tonnara of Arenella, near Palermo, and took leases on the neighboring tonnaras of Vergine Maria, San Nicolo, and Solanto. Further west, in the province of Tràpani, he leased the tuna works of Favignana, Formica, Scopello, and San Giuliano. Immediately the thirty-year-old entrepreneur began to educate himself, becoming an attentive disciple of the raises. He observed, listened, then went into action, reorganizing and reviving Sicilian tonnaras. He must have been a man of considerable charisma to have been able to convince the raises to change their ways when for millennia no one had been able to do so. Florio streamlined maneuvers, redesigned the traps, and sometimes changed their placement at sea, bringing record catches—as much as three thousand tuna in a single day.

In the canneries he made use of previously discarded tuna parts, pressing oil for wood finishes from inedible bits and grinding the skeletons into a fine powder prized as a fertilizer. Later owners of the Favignana tonnara followed his lead and made handballs from the giant tuna eyes and, in the 1920s, even experimented with a perfume made from bluefin blood. Florio found a way to pack tuna in olive oil for mass consumption and opened the way to worldwide exportation of his product.

Despite this success, Vincenzo Florio, for some unknown reason, did not renew his lease on the tonnaras of Favignana and the tiny nearby island of Formica. In 1859 Pallavicino's heirs leased them instead to Giulio Drago. When Drago took over he didn't have enough capital to prepare the ropes and cables for the trap and so was forced to sell the tuna at a measly twelve lire each before they were caught to raise money. But in just a few years Favignana's tonnara made Drago so rich he was able to give the daughters of his tonnaroti the huge dowry of three hundred lire each so they could marry well.

When the last of the Pallavicinos died, the heir decided to sell the Egadi Islands. Ignazio Florio, Vincenzo's son, rushed in to buy the islands before his competitors could make up their minds. On March 7, 1874, the Egadi Islands, including the tonnaras of Favignana and Formica, became Florio property for the sum of 2.7 million lire. Ignazio ran it with his son, Vincenzo, named for his paternal grandfather.

Ignazio proved as good a manager as his father had been. Under his hand Favignana flourished as never before. A sandstone tablet on the cannery wall proclaims that in 1878 the tonnaroti killed 10,159 tuna. The Florios also invested in tonnaras in the Canary Islands and a new one in Tunisia in partnership with the Parodi brothers of Genoa.

The Florios' fabulous wealth and magnanimity brought good times to Favignana. Everyone had work; in fact, cannery employees had to be imported from Marsala and Tràpani. Florio paved the town's pounded dirt piazza with marble slabs. On the ruins of a Moorish fort he built a Belle Epoque palazzo for his family's spring visits. Favignana had electricity before Rome did; Florio needed it for his cannery. He built servants' quarters, carriage houses, and stables on the rise above the port. Cooks prepared meals in a kitchen behind the stables and brought them into the mansion through an underground tunnel because Florio couldn't stand the smell of food cooking. He built a brine tank where gill net fishermen could drop their catch of the day. If they were lucky, Florio's chef chose their fish for his master's dinner. It was a time in Sicily when peasants still identified with the wealth and prestige of their overlord. A mighty baron, or businessman, would take care of them.

Florio loved Favignana and wanted to make its tuna works the queen of tonnaras. It was already rich; he wanted to make it beautiful. He hired one of Sicily's best architects, Giuseppe Almeyda, who designed Palermo's elegant Teatro Politeama, to design the Camparia and cannery. Their soaring interior spaces reflect Palermo's Norman-Arab sacred architecture.

Because of Florio, what had been a sleepy medieval village now had international renown. His was the largest cannery in Europe. The prized, tender belly meat of the bluefin, packed in olive oil, was sold in bright yellow and red tins decorated with emblems of prizes won at international trade fairs: the Diploma of Honor at the Exposition of London in 1888; two more from the Palermo exposition; the Grand Diploma of Honor at the Italian American exposition at Genoa in 1892; the grand prize at the Milan fair of 1906; and medals of honor from Turin, Berlin, and Milan. Favignana's name appeared on every can.

The islanders loved Ignazio Florio. They raised a statue to him in front of the town hall in Piazza Europa. Inscribed on the pedestal are these words:

Out of civic honor
the workers
with money from their own pockets,
and the town government
in official agreement,
raise this monument
to the benefactor
of
Favignana
to show
that discord need not exist
between Capital and Labor
between Wealth and Poverty
Where there preside
Justice, Mercy and Love

The islanders hired a famous local sculptor to carve into the pedestal the intertwined symbols of the Florios' great wealth: the caduseus for pharmaceuticals; a cornucopia spilling coins for banking; a tonnaroto's gaff for the tonnaras; a halberd (still used by tuna butchers) for the canneries; grapes for the Florio vineyards; barrels for their Marsala wine; a ship's wheel for their cargo fleet; anchors and chains for their foundries; and a compass and map for their vast land holdings.

Half a block from the statue is Zu Isidoro Marino's bike shop. Zu Isidoro retired after a life inside La Florio, the islanders' name for the cannery. His father and grandfathers worked there too.

"Ignazio wanted nothing but the best for the Favignana tonnara," Isidoro said. In those days there was a mattanza every day in the spring, sometimes two. Florio added his private boat to the traditional tonnara fleet. He had it built of expensive tropical wood, put a Persian carpet on the floor and a canopied red velvet seat at the rudder, then hired ten men to row him to the trap to witness the mattanza, Zu Isidoro said. Florio had the tonnara's two towboats painted yellow above and blue below—blue for the bluefin in the blue deep, and yellow, the color of gold, for the luchre they became once lifted from the sea.

According to one account, Ignazio, on his deathbed, told Vincenzo to "sell everything if you must, but keep the Favignana tonnara." But Vincenzo could not honor his father's wish because the family fortune fell on hard times.

In 1937 the Parodis, the Florios' partners in banking and the Canary Island tonnara, bought Favignana's tuna works and the prestigious Florio trademark.

"When the Parodis ran things, money hit you in the face," said Rosario Ritunno, at work twisting rope in the Camparia courtyard. He looked up at the empty third-floor windows of the Parodi mansion next door. "One year we took 699 tuna, but Parodi said, 'Let it count for 700,' and every man got an extra 50 lire in his bonus."

"The Parodis used to cut up a tuna without weighing it and give a piece to each family," Giuseppe Aiello, the gateman, said.

Like the Florios before them, the Parodis came to Favignana every spring for the tonnara. They flew their family crest from the boathouse roof when they were in residence. "The signora came with twenty servants to open the house," Zu Isidoro said. The Parodis personally managed the tonnara until 1985, when they leased it to Franco Castiglione, son of Nino Castiglione, a former tonnaroto, now a self-made millionaire of Tràpani.

Now the Parodi palace looked haunted, the garden dry and overrun with weeds, and when Luigi Parodi comes in the spring he comes alone and takes a small room at the Four Roses campground.

7

The Cannery

These days, when the tuna were dead and massed in the longboat, the tonnaroti covered them with ice and a tarp and towed them to Tràpani. At the city's harbor a crane lifted them into trucks; they dribbled blood on the road all the way to Castiglione's slaughterhouse, where they were counted and weighed. They were laid out on a cement floor, beheaded by Sicilians swinging halberds, and slit up their ventral side by young Japanese tuna butchers with just-sharpened knives. The butchers plunged their hands up past their wrists into the visceral cavities and drew out long sperm and egg sacs, which they tossed into red plastic buckets.

By this time, back on the island, all that would be left of the tuna were scarred hands and stained T-shirts, the spattered blood already turned brown.

In the past the tuna came back to Favignana. The islanders grouped on the shoreline to welcome the bountiful catch, to see their colors. I've seen an old film of a 1957 mattanza. There was a Japanese businessman there even then. After the fishermen tie ribbons around the crescent tails of the tuna he has chosen, a strong man ties a rope around the base of the tail, turns around, and hoists it onto his back. He staggers across the strand to a platform where it is weighed and put in a wheelbarrow. The fish passes through a gate in the walls of the Florio cannery to be cooked and conserved. When the tuna left Favignana, it was in a can with the island's name on it.

Zu Isidoro told me that once, during World War II, when the island was being bombed, the cannery director told the rais to set the trap, but he didn't inform the Parodis of the risk they were taking; all the men, boats, and equipment could have been blasted to bits. But the gamble paid off. They took seven thousand tuna that year, and there was work at the cannery, Isidoro said, and food for the Favignanesi.

The cannery's closure in 1981 came as a disaster for the island. A thousand people had worked there—fish butchers, mechanics, stonemasons, stock boys, carpenters, cooks, carriers, crate makers, clerks, cashiers, day-care workers, plumbers, electricians, secretaries, guards, janitors, can makers, packers, and shippers. La Florio was always hiring help. Favignana alone couldn't handle the demand for workers. In Sicily, where a job is so precious it's against the law to hold two, even the island women worked, without social censure. People ferried over from Tràpani and Marsala, and Zu Isidoro said Florio had prisoners brought to the plant in chains. When it was not tuna season the cannery packed sardines and fish brought by trawlers from all over Sicily and from as far away as Naples.

Now only one employee was left, Vito Giangrasso's father, Peppe Nue, the custodian. One morning I bicycled to my appointment with him at 7:00 A.M., the hour he reported for work every day, just as he had done when the whole island once showed up for work there. Skeleton key in hand, he stood at the crack between the fifteen-foot-high double doors and waited until I had locked my bike and found my notebook. The doors were set in a tufa wall twenty feet high that surrounded the plant's thirty-five thousand square meters of workspace like the wall around an ancient city. The golden stone smokestacks towered above it like temple columns.

Peppe Nue stood at attention, blinking in the sun that baked the cream-colored stone. When I was ready he turned the key in the lock and, with a Sicilian sense of drama, pulled the doors open slowly, majestically. I sucked in my breath at what I saw.

Another world, a cool shady courtyard, an oasis of green, smelling of soil and pine. Pines that were thirty feet tall, thick-trunked and fragrant, the tallest trees on the island here inside these high walls. A dripping, hanging ficus tree imparted a languorous splendor. There was a feeling of peace and protection, as in a secret garden. This was the scene that once greeted the men and women of Favignana when they came to work every morning.

"As soon as I open the great door, I see what it used to be, and I feel like I could cry," Peppe Nue said.

For the next two hours he walked me through the plant that had allowed twice as many people to live on the island. The cannery gave them pride of place. The fish came from their waters. The people knew that what they made here was the best in the world.

The rectangular yellow tins contained the highly prized *ventresca*, the belly meat, while the green tins contained the *tarantello* from the tuna's flanks. Both bore a fine engraving of a lion on a triangular island crouching to leap from the sea, an image taken from the Florio coat of arms. Two bluefin tuna swim above two anchors and over the words "True product of the tonnara."

Peppe Nue had worked for the Parodi family, who bought the Florios' tonnara and brand name in 1939. But the Parodis sold the plant to the regional government, which planned to make a museum of it, and so the region became Peppe Nue's new boss. He guarded a memory now. The place was left just as it was on the last day of work, a dusty still-life, a sad shadow of its former grandeur.

The workers had entered the gate and walked through this verdant courtyard before they punched in at the pay booth. The hours were posted plainly: 8:30 A.M. to noon, and 1:30 to 5:30 P.M.

My greatest impression was of the respect and dignity Florio accorded his employees, as witnessed by the workplace he gave them. The working women could drop off their babies at the free nursery where young women cared for them in shifts. There was a playground, a slide, a sandbox, short wooden stools and low tables, and white enamel chamber pots with handles like old-fashioned coffee cups. At ten in the morning and three in the afternoon, women with suckling children could come for half an hour to nurse their babes.

Next to the nursery, in this same courtyard, were the wooden quarters of Florio's two giant German shepherds, guard dogs that were loosed at night to bark at intruders. We followed the path the workers took, under rounded and pointed archways. The windows were shuttered and barred with ornate iron grilles, each bearing a foot-tall letter "F" in a circle in the center. Peppe Nue said Florio loved beaten iron. A statue of a smith pounding an anvil used to stand at the entrance, but the Florios took it with them when they sold the cannery to the Parodis.

We passed the carpentry shop, where men and women made shipping crates. A sign painted on the wall sounded like one of Mussolini's slogans: "L'Industria domina la Forza" (Industry rules Strength). We walked through a warehouse of carved wooden cabinets, ornate armoires almost eight feet tall. Inside their locked doors was the cannery's inventory—everything from replacement motors to typing paper. The regional Department of Culture had taken possession and inventoried most of the items and tagged everything there. Otherwise, nothing was disturbed. It was a poignant scene: the tools of the tonnara, lying there like props brought to a "come as you are" party and then left there.

There were great wooden sea chests, tied with rope, mounded up in the middle of the gloomy room. The stones in the hallway were green with moss. There were changing rooms for men and women, a storeroom, and a lonely statue of the Virgin Mary at the entrance to the *galleria*, where the filled tins were capped. Then there was the *batteria*, a long line of copper cauldrons where the tuna had been cooked over charcoal fires.

"We used the *carbone amaro*," Peppe Nue said—the hard bitter coal, not the sweet.

We went into the outer room called *la giungla* (the jungle) because of the ropes that hung down like thick vines in a rain forest, where the beheaded tuna were hung to bleed. The room gives on to the sea, and the iron bars of a gate cut it in vertical strips. The tonnaroti used to pole their heavy boats up to this shallow port and the *massari* (massive porters) would load the tuna, often a thousand pounds, on their backs and carry them here to be beheaded and hung.

We walked through the machine shop where the cans were made and the print shop where the tins were stamped. Two stacks of nested containers leaned up from the floor like blossomless lily stems. Soft green light filtered through tinted skylights into the long arched corridor where the boiled tuna had been canned. It too had the aura of a cathedral. Even the boiler room and the freezer had an air of noble magnificence about them.

Peppe Nue showed me where masons made cement, the janitors' courtyard, the ice factory, and the smithy where the anchors were made and repaired. A dusty brown bottle of brandy stood on a thick wooden desk. The door to the smithy had a tiny bud vase lashed to its screen. A long-stemmed carnation whose brittle brown corpse had neither crumbled nor lost all its petals leaned in a graceful arc from a vase gone dry long ago.

Nothing was wasted here. La Florio made use of the bluefin's meat, eyes, blood, oil, and bones. The last thing Peppe Nue showed me was the Campo Santo, the holy field, which is the walled tuna cemetery where the bluefin skeletons were dried before being ground into fertilizer. Even this cemetery gave life.

Zu Isidoro told me that during the war, when he was a hungry boy, he had scaled the Campo Santo's wall and sucked the marrow from tuna bones.

Sushi

When a Japanese businessman takes his important client into one of Tokyo's thousands of *sushi ya san* (honorable sushi shops) the men behind the counter yell a welcome. Not a polite hello, but a forced warrior's shout. The businessman's firm is about to pay seventy-five dollars for two bites of the fresh, red belly meat of the bluefin, served on a pure white plate. Two ounces, seventy-five dollars. The atmosphere is cheery, bright, and male.

I have never been to Japan, but my friend Judy Anton, who lived there for many years, knows all about the sushi ya san. She spent much of her childhood in Japan, where her father taught at Tokyo and Kyoto Universities. Judy, a woman of many talents, signed her first jazz recording contract in Japan, was a model for Japanese cosmetics, hosted her own late-night talk show on Japanese television, was a consultant for American businessmen doing business in Japan, and now designs art quilts using exquisite hand-dyed fabrics.

"The sushi ya san treats its customers like gold. Everybody thinks they've found the world's best sushi bar," she said. "Everybody in a sushi ya san is treated like a king. That is one reason to go to one of these. . . . Only men prepare sushi in a sushi shop. It is said the oil from a woman's hands ruins the sushi."

The businessman and his client order *toro*, the meat that runs in a diagonal stripe across the belly of the bluefin. In Japan toro is the bluefin meat with the highest status.

Even the toro meat is graded by status: the more fat, the more prestige. *Maguro* is almost fatless, *chu-toro* is medium-fat, and *jòtoro* is high-fat. Jòtoro is the most prestigious, according to Judy, but some Japanese don't like it because it coats your tongue with oil. "I think a lot of people like it because it is the most expensive. If people are willing to buy it for you it shows great respect."

Judy described a typical scene after the men sit down to order. The businessman host asks his guest, "What would you like?" The guest replies, "Chu-toro." The host offers jòtoro. The guest declines, saying, "No, I like half-fat, chu-toro." Then the host is at pains to make sure his guest is not just being polite.

"You pick the most expensive toro you can afford," Judy said. "In Japan it is not so much what you like that matters. They are very, very intent on showing where you stand in society—the kimono you wear, what fabric it is made of, how you style your hair. If you have an American car, that is the height of all heights. *Bella figura* is very important in Japan. If bluefin tuna is all the rage in Japan, you must know that and offer it to guests. The host of the sushi bar, if he knows you can't afford it, won't offer it. Most customers call ahead and pre-order, and the sushi ya san host will offer that to you when you walk in, as if you hadn't had the telephone conversation."

The Japanese have old values, Judy said. "Everything you can think of is geared toward status."

I'd heard that Japanese connoisseurs could tell by its taste how frightened a tuna was when it died, or even how it was killed. Judy said that story was "probably untrue."

"It's like talking about wine," she said. "The more detailed you can be about connoisseurship, the more status there is."

Japan is the world's largest consumer of tuna. The bluefin the businessman orders most probably comes from the Tsukiji market, known as "Tokyo's kitchen," where the average price of a bluefin sold at the daily 5:30 A.M. auction is $20,000, with the rare giant selling for three times as much. Fishmongers bid on the tuna, then carry their purchases back to their market stalls, where they slice them up carefully and sell them to restaurants and sushi bar chefs.

Lucille Craft, writing for *International Wildlife* magazine, interviewed an eighty-four-year-old fishmonger at the Tsukiji market. Kozaburo Omura, whose shop dates back fifteen generations, remembers when the Japanese did not like the toro, now the most prized cut of bluefin. Before the Second World War the fatty belly meat was thrown out. But after the war General MacArthur's staff said that discarding the belly meat was wasteful. Then people started to eat toro.

In the mid-1970s, when refrigeration technology became sophisticated enough to freeze the meat for shipping without spoiling its delicate taste, toro became the ultimate in *sashimi*, sliced raw fish served without rice. Toro became the Japanese equivalent of caviar. People began to pay incredible prices to get it.

On January 9, 1992, one giant bluefin weighing 715 pounds sold for $83,500, according to Carl Safina, director of the Living Oceans Program of the National Audubon Society. That's $117 per pound. The fish would provide 2,400 sushi servings and bring in $180,000 at $75 a plate.

In the 1970s, when fish distributors realized they could air-freight whole, fresh bluefin to Japan to capitalize on the huge demand for sushi and sashimi, more fishermen targeted bluefin weighing anywhere from three hundred pounds to a half-ton. Prized jumbo bluefin caught off the New England coast were placed in coffins on shaved ice, covered with a golden space blanket, nailed in, and sent off on a jet plane to Tokyo, where they were auctioned off at the Tsukiji market just seventy-two hours after they had been caught.

The extraordinary power of the Japanese fishing lobby has even slowed down that country's space program. Because Japan's rocket launch site is near one of the country's richest fishing grounds, airborne rockets drop debris and spent stages into schools of Pacific tuna, infuriating

Japanese fishermen, who have demanded that launches be limited to the months of February and August.

In 1990 the Japanese fishing fleet was the world's largest in terms of tonnage, number of vessels, and number of landings. Nonetheless, Japan's tuna imports—already at $1 billion in 1991—have skyrocketed in recent years. As much as 35 percent of domestic consumption of raw tuna, the most important segment of the Japanese market, is now imported, according to a U.S. National Marine Fisheries Service memorandum. This despite the fact that it takes a long-haul tuna boat fifty days to sail from Las Palmas in the Canary Islands off the west coast of Africa to Japan, at a cost of $7,000 a day.

In the South Pacific, off Satawal Island in the Federated States of Micronesia, where modern tuna boats may catch 150,000 tuna in a single set of the net, island fishermen using pearl-shell lures and canoes still hunt tuna in the traditional way. The anthropologist Robert Gillett studied their traditions for more than a decade and in 1987 published his findings in the *Bishop Museum Bulletin in Anthropology I*. To the islanders, he reported, "tuna fishing is . . . fun. . . . Even when catches are poor, hardly enough to justify a trip for the food value, there are no crew shortages. Loss of sleep, bone-chilling rain, baking sun, and hours of monotonous transit to and from the fishing ground aboard a pitching, rolling, jerking canoe are considered small sacrifices for the thrill of polling tuna," he wrote.

The islanders' hunt was saturated with magic and taboo. On Satawal the tuna were regarded as sacred, and the use of their ordinary name was often forbidden. When a returning fishing crew could not see the island of Satawal because it was obscured by a bank of clouds, nearly all the crew began making "X" signs with their hands toward the horizon to displace the clouds, Gillett wrote. When one canoe had been caught in a series of nautical mishaps and its catch rate was low, "the ghost thought to be causing the problem was noisily chased away by dozens of people."

On Satawal taboo prohibited women from involvement in tuna fishing; men could not have sex for a month before fishing, nor could they eat coconut crab, big-leaf taro, or bananas. Taboo also prohibited making coconut fiber twine during the fishing season; fishing by anyone who stepped on excrement before or during the launching of the canoe; possessing metal on the fishing grounds; eating part of the tuna catch on the fishing ground; the eating of tuna entrails by anyone except old men; having a hooked fish fall back into the water; and opening the fishing gear box without first cleansing the hands with the leaves of a certain tree.

Gillett's report included some findings of H. Hijikata, a Japanese anthropologist who had lived on Satawal for seven years in the 1930s, when taboos were even more intense: Skipjack tuna could never be brought into the house; when not all the fish could be consumed at once the remainder had to be placed in a basket and hung outside on a tree or other support; the flesh was eaten with the hands, with no salt or seasoning. No one could eat tuna while standing up or walking around. Those who had eaten the skipjack could not go into the sea for the whole of that day, and if they washed they had to use freshwater. The bamboo spatula used in cutting up the fish could not be used for any other purpose. Women in the menstrual hut could not eat or even touch skipjack. Those coming out of menstrual isolation could not eat skipjack until seven nights had passed. Anyone who ate bananas could not eat skipjack for three months. The tail and the attached bones could not be discarded but had to be tossed up onto the roof or hung in a high place.

The bluefin that swam unawares into the Favignana trap were fat and delicious. It was only the insatiable appetite of the Japanese for bluefin that kept the Favignana tonnara afloat in recent years. The Japanese waited with sharp knives at Castiglione's slaughterhouse for the Chamber of Death to give up its fruit.

The tonnaroti of Favignana could taste the bluefin only in their memories. They tried to forget the aroma of cooked bluefin pouring from the cannery's stacks and waited in the sun for the trap to fill again.

These days the bluefin steaks on Favignana come from Tràpani. One of the best kitchens on the island belongs to the Guccione sisters, Maria and Giovanna, who run the Albergo Egadi just off the main piazza in Via Colombo. Chef Maria gave me her recipe for tuna with mint, a dish redolent of the aromas of Morocco and Tunisia. "With this recipe, we recall the Arab origins of Sicily," she said.

Cut fresh tuna steak slices about three-quarters of an inch thick. "Not too fatty," Maria advised. Let them soak in red wine a few minutes, then coat with flour. Arrange the tuna steaks in a buttered pan. Place fresh mint leaves on top. "Just a few." Add finely chopped onion and a bit of chopped parsley. Cover the tuna with a layer of potatoes sliced as thin as possible. "They should become a cream in the cooking," Maria said. Add salt and pepper to taste. Pour in the wine in which the tuna was soaked, add just a little water (*pochissima*), cover, and cook over a very low flame.

"Don't let it stick," Maria warned. Continually shake the pan. Cook for twenty to twenty-five minutes, until the potatoes become creamy.

"This is a recipe from my childhood," Maria said. "We serve dishes with a history."

The Anchors

At the Three Crosses intersection at the western edge of town, where one road slopes down to the shore, a most unusual painting of a madonna looks out to sea. She is the Madonna of the Tonnaroti, a Sicilian beauty of the Arab type, black eyes, aquiline nose, with a wisp of black hair escaping the blue veil she draws close about her face and shoulders. She wears a real gold hoop earring in her right ear and a sapphire ring on her finger, both embedded in the canvas. The Three Crosses of Calvary top the tufa monument that frames her portrait. Above her is the crucified Christ, who gave his life so that men could be saved. His cross faces the long straight road to Punta Lunga. But the madonna stares in the opposite direction, down a street, over a strip of beach laden with anchors, and out to sea.

In her arm she cradles a tuna instead of the Christ child. An artist who recently restored the painting said that its style is typical of the 1700s, and that tonnaroti commissioned the work.

Rosetta Messina, a widow who lived a block from the madonna, said the painting was not commissioned but found. The fish became part of the picture only when, one year when the catch was poor, one of the Florios brought the Virgin a tuna as an offering and she took it in her arms. The tuna's eye is a bright blue turquoise bead.

In her right hand the Virgin holds an old-fashioned three-pronged anchor, precursor to the hundreds lined up on the beach below her. Soon they'll be attached to the steel cable skeleton of the trap. They will dig into the seafloor and strain against the eastern current that tugs at the nets and threatens to rip them apart at the seams; they will keep the trap whole and the barrier nets still. Nearly four hundred anchors stand on their sides in long lines on the strand, aging giants forged a century and a half ago. The smallest of these are 6 feet long and weigh 660 pounds; the largest ones surpass 10 feet and weigh nearly 4,000 pounds. The beach will be unburdened all spring while they are at their work.

Once they had a black patina; they were scraped clean and painted with pitch to protect them from the salt air. But now they corrode on the beach all winter, getting crusty, brittle, and a little thinner every year. Yet they face the sea with the pent-up energy of an arrow set in a bow.

The largest ones have names: Levante—The East, the keystone anchor, weighs 3,740 pounds; Marsala always takes the position closest to that city on the main island; Tramonto—Sunset will lie farthest to the west; and Santa Caterina will lie at the base of the mountain.

Such anchors lie on the shores of extinct tonnaras all around Sicily, abandoned as they were the last time they came out of the water. I have seen them at Acqua Santa, Vergine Maria, and Scopello, forlorn marine menhirs lined up on the beach; someday someone will wonder what they were and why so many.

But at Favignana the anchors are protected by the Virgin, and the Virgin is cared for by the women. It is the only way women still participate in the tonnara. Years before they helped make and repair the nets, seated on wooden chairs in small groups before their doors in the twilight in the stone-paved streets. Now the women take care of the Virgin who takes care of the tonnara.

The tonnaroti themselves never come to this image of Mary but revere the Madonna del Rosario, kept in the church of Sant'Anna in the oldest part of town. Their wives and mothers tend the Madonna of the Tonnaroti and keep the vases set before her filled with fresh flowers.

A wooden dagger hovers above the Virgin's right shoulder, and although Padre Damiani, a priest and local historian, denies it, Rosetta Messina says it is the kind of blade a Sicilian sorceress would inscribe with prayers and place on a velvet pillow to ward off evil. Until recently new brides of the *quartiere* would visit the shrine on their wedding day to ask for a happy marriage and to drape a white gold bracelet or necklace on the dark madonna, who would wear it until it was stolen.

The madonna's gaze is trained on the trap site. Her view of it must remain unobstructed, Rosetta Messina told me.

"You cannot build anything in front of that madonna," she said.

"Why? Is there a law?" I asked.

"No," she said. "Any building that blocks her view will collapse. The Madonna will make it fall." She said the image was miraculous. "She's been stolen several times. People from the forest on the other side of the mountain have taken her. And the next morning she always returned, by her own power. Once they even walled her in, and she still escaped."

Now the anchors were being loaded, sixteen to a boat. Buonavventura Torrente drove the forklift that carried them one at a time to a block and tackle at the shore's edge. He lowered an anchor to the ground and drove off through a gate in the sandstone wall to get another. A handful of men attached chains to the anchor, which was then hoisted aloft, swung around, and suspended over a waiting workboat. One man operated the crane while seven others cajoled the anchor into position above the boat, maneuvering it so that when lowered it would fit neatly, like an overlapping feather, into the orderly array of already loaded anchors. The anchors' exact placement, layered eight on each side of the boat, was crucial to the next operation—*u cruciatu*, when they would be dropped at sea. One anchor could not be allowed to pull another from its place when it was pried upright and tipped overboard, or some unsuspecting tonnaroto could go with it.

Fafarello loved the anchors. A short man, close to retirement, he had blue-gray eyes and close-cropped silvery hair under a blue felt beret and wore loose, faded blue work pants and a blue smock. He rarely spoke, but something sang inside of him. The sun and salt air had creased his face into a permanent grin around his broad nose. He walked, quick and nimble, along the gunwale to where the anchor was suspended on its side. He spread his short arms along the anchor's curve, pressing his chest against the giant, commanding it like a Lilliputian elephant tamer. The younger men called out warnings to him: "Peppe, stai attento!" Twice the lumbering giants had injured him. Once a rusted anchor split in two as it was being dumped into the sea, hit him in the chest, and put him in the hospital for the entire tonnara season.

The motorized hoist was first used in the mid-1960s; until then, incredibly, the anchors were loaded by hand. One day in March the rais would tell the men to come back from their lunch break in shorts. Thirty men would slide a couple of thick wooden beams under Levante, pick her up, and carry her into the frigid March sea. When they were chest-high in the water they would lift the anchor into the boat. The smaller anchors required only fourteen men to bear each one.

Once aboard, the anchors looked graceful and light. Sixteen oaken stocks, angled toward the prow, rested on a long pine plank, like ribs along a spine. The anchor crowns hung over the side of the boat, flukes pointed toward the sky. When Fafarello poled through the shallow water to the dock, the anchor boat looked regal, like a blue swan with black wings.

Rosetta

Seventeen is an unlucky number in Sicily; travelers cancel airline reservations on the seventeenth of any month, and high-rises here have no seventeenth floor. When the tonnaroti measure a net with a meter stick, they count aloud: "sixteen, sixteen-and-one, eighteen." Not surprisingly, Saint Patrick's Day, March 17, passes without a parade. But Saint Joseph, whose feast is two days later, is adored throughout Sicily.

Saint Joseph is a blue-collar saint very popular on the island; he is the protector of carpenters and orphans. His feast coincides with the vernal equinox and is celebrated in disparate, sometimes pagan ways. Residents of Palermo light three-story bonfires in the streets and burn old furniture. On the island of Marettimo fireworks are set off, in Campobello di Mazara altars of bread are made, and in Siracusa the oldest boat is burned. Saint Joseph walks through the town of Salemi knocking on doors looking for food for his family and a place to stay and is turned away three times. It is a ritual in which the whole town participates. Each person knows his lines.

"We are three poor travelers, we are tired of walking. Lodging we would like for this evening."

"There is neither room nor food. Get on with you to some other place."

Sometimes Saint Joseph is accompanied by a drummer and a narrator who explains the story to the crowds: "Saint Joseph walked from his friends to his relatives, but no one was there to do anything for him at all."

Pathos rises. In the second phase of the ritual the townspeople recognize his family tie to divinity and invite him to a feast. In Sicily it's always who you know that counts.

One year I was in Favignana for Saint Joseph's procession, which usually takes place about the time the anchors are dropped. As I biked from my house at Bue Marino to town I heard the raucous blare of a brass band. I followed the horns to the solemn parade. The town band, dressed in maroon blazers and black pants, was marching through the back streets, followed by two girls handing out bread rolls. The elderly followed in cars, peering through rain-splattered windshields.

A girl handed me a roll, and I ate it on the spot. But Antonio Casablanca later told me that seamen save the *panuzzo* for days when the sea is rough. They toss the bread into the water with a prayer to calm the waves so they can work. He gave me his to save, and now, five years later, it was hard as a rock.

Saint Joseph's statue did not join the procession. That year his feast fell during Lent, the forty days before the Crucifixion when Catholics are supposed to fast and focus on Christ's self-sacrifice, so the saints had to stay in their church niches, covered with purple shrouds. I followed the band into the piazza, where a family dressed as Joseph, Mary, and Jesus sat on a raised stage at a table set with a decanter of red wine. A church lady poured water from a pitcher into a bowl in which the Holy Family washed their hands. Onlookers packed the piazza. Cristina's father sat at a table outside the Two Columns bar; he waved me over to the seat beside him.

"Better dead in Favignana than alive in Torino," he said, apropos of nothing. His sister-in-law in Torino wanted him there for a few months, he said, but he couldn't breathe in the city. "Signorina, the air here is different." He called me "Signorina" because he couldn't remember my name.

Then he threw me a curve. "Do you know that a headless corpse can walk?" I didn't. He pointed his chin to a spot up the street, in front of the bank. He had seen it himself, during the war, a man's head blown off by a bomb. The decapitated corpse stepped off the curb as if to cross the street, he said, then crumpled in a heap.

Antonio Casablanca, one of the younger tonnaroti, joined us for coffee. Every year his aunt, the widow Rosetta Messina, transformed her sitting room into an altar to Saint Joseph for the feast, one of the last people to keep the old tradition. Today was her father's name day, and Antonio wanted to wish his Zu Peppe a *buon onomastico*. He invited me along.

The sky had cleared, and now the sun was hot. Zu Peppe, an old tonnaroto, sat on a wooden chair in his front yard in the shade of an unkempt bush, his swollen, ulcerated left leg bandaged and propped on a stool. He was of a breed now nearly extinct. The one-story house behind him was crumbling, the plaster cracked and chipped. He scolded Antonio for not coming more often. Rosetta appeared in the doorway and wiped her hands on her apron.

"He's having a bad day," she said. Her dark brown hair was shot through with gray and pulled back under a headband.

"Last year, at age ninety, he escaped from the house," Antonio told me while his uncle half-listened. "He went up the mountain, in winter, by himself, in his slippers and pajamas, went halfway up the mountain to a ruined old house. Three days he was gone. I spent two days with the police, helicopters, and dogs sent to search for him. Divers looked for his body in the sea. On the third day he came down by himself, all torn and bloodied because he had fallen on the rocks. He thought it was the war and the soldiers were after him and that his house was about to be bombed."

When Zu Peppe reappeared after having outwitted the police, the police charged Rosetta with neglect and presented her with the bill for the search. It ran into the tens of thousands of dollars, she said. She seemed unruffled about it; the case would not come to court for years. She beckoned me to the door.

Rosetta's altar was a phantasmagoria. Great hanging swaths of powder-blue satin pinned with cardboard angels receded in layers to the back wall of her sitting room. Some of the angels were store-bought cherub heads with rosy cheeks, blond curly hair, and wings sticking out of their necks. The others were seraphim she had drawn, colored, and cut out a long time ago. She had strung silver garland along the curtain rods and hung them with Christmas tree ornaments. The satin curtains framed a portrait of the Holy Family, at whose feet stood silver candelabras set with long white tapers. Between them stood ceramic vases filled with fresh gladioli and plastic carnations.

Her front door was open, as it was only for wakes and feasts; she was letting the saint's blessings radiate from her altar. Rosetta pulled up a chair and told me to sit and look, if I liked. She pointed out the piece of satin that was stained by the rotten banana Zu Peppe had stowed in the linen cupboard and forgotten. She always left the altar up until March 25, a feast of the Virgin, she didn't know which one. (I looked it up—the Annunciation.) She told me I could come back and take pictures.

I did. We became friends, and so I was often at her kitchen table for coffee or a meal. A pockmark in the ceiling above our heads reminded her of the time the espresso pot exploded and embedded itself there. Throughout the house the plaster crumbled and fell in a fine dust. "Every day I have to wipe the ceiling off the floor," she said.

Rosetta was sixty and cared for her failing father, her bedridden mother, who no longer recognized her and thought she was a thief, and her older sister, Filippina, who was mentally handicapped and barely able to walk. The room next to the altar, which Rosetta shared with all three, looked like a hospital ward.

Her niece, Vita, was always there, in an apron, in the kitchen. Vita lived with her new husband, Mimmo, in a house her aunt had given her, but she came daily to help with the family's meals. This was Rosetta's mother's house, where Rosetta and her siblings had grown up. But her brothers and sisters had all moved away, so she nursed her father, administered injections, changed his bandages, dressed her sister, fed her mother, changed her diaper, bathed her, and propped her up on clean linen. Sometimes her mother refused to open her mouth to eat, and Rosetta would rage and scream because "if Mama doesn't eat, she'll die."

Rosetta saw visions. People appeared to her in clouds in the kitchen and spoke to her while she was knitting. The visions often gave warnings. She wore a gold locket with a photo of her dead husband on a gold chain around her neck. She had married late but happily. "He held me like this," Rosetta said, cupping her palm as if she had been a jewel protected in the middle of it. He bought her a bathing suit, he took her to the beach. After just nine months of marriage her husband had died of heart disease.

It was hard to think of Rosetta as one who liked to be protected; she was short, saucy, and loud. On days when her knee wasn't killing her she'd burst out in song while she hung the wash. She was a widow, but she didn't wear black because she was too alive. She lived off the government pensions and sickness subsidies of her mother, father, and sister.

Mimmo, Vita's husband, made himself useful around the house. He had built a new bathroom and spackled the walls. In the summer he was a pastry chef at l'Approdo D'Ulysse, an island resort, but in the spring he did construction work. When he came home for lunch every day Vita planted a Hollywood kiss on him as he entered the kitchen. They were still newlyweds.

Vita would seat everyone while Mimmo washed. When he took his place at the head of the table he would wipe the drool from Filippina's chin and tease her. Filippina had a crush on Mimmo, the only one who could make her smile. We always ate on a clean tablecloth: bowls of black olives, pasta with fava beans in red sauce, cakes of fresh sheep ricotta delivered by the farmer himself, a salad of tomatoes, onions, and capers, and chewy bread that Mimmo made.

On weekends he and Vita scoured the beaches for driftwood to fire Rosetta's brick oven. She kept an outside kitchen, as many Sicilian women do—stacks of dishes on shelves, a sink, a two-burner stove, strings of garlic and dried red peppers on the wall, pots of basil, cases of mineral water in plastic bottles, a cabinet lined with jars of baby food for her mother. A plastic roof cast a green pall on this pocket courtyard; the shadows of cats padded overhead. In summer it was a sauna, but in winter it stayed dry. Rosetta's brick oven was in the corner.

On Saturday evenings in the spring Mimmo made bread for the week. He measured by eye, mixed the dough in a plastic tub, kneaded it, and let it rise. Vita lit the fire; the flames licked the bricks. When the coals were hot and the ashes just right she slid the round loaves into the dome on a long-handled paddle. When the first loaf had cooled, Rosetta carved a cross on its undercrust. She held the loaf to her breast and carved inward with a kitchen knife. She drizzled the first slice with olive oil and handed it to me.

With the leftover dough Mimmo made pizzas, and then the kitchen filled with cousins who just happened to drop by.

11

La Costa Alta

The rais was polite but distant, busy and very nervous. Inside the courtyard there was much cutting of twine and twisting of rope, laughter and loud words. I observed the men silently, still feeling very much the outsider. If I did not speak to the tonnaroti, they did not speak to me.

Their simple, ancient hierarchy went like this: The rais was king, general, and pope. He had two lieutenants with the title of capoguardia. These were seventy-four-year-old Filippo Messina and Occhiuzzi, the rais's blue-eyed cousin. Under them were the six boat captains, lords only in their own vessels. Everyone answered directly to the rais.

Occhiuzzi had another duty: He was the *Prima Voce*, the ciurma's "First Voice," or soloist, who sang the verses of the traditional work songs during the mattanza. Occhiuzzi was the last working tonnaroto to know the words by heart. Once the Prima Voce also sang on land to give the men rhythm and strength.

"What will happen when you're gone, Occhiuzzi?"

"Then we'll have a silent mattanza," he said. No magic, no supplication, no threats to God. That's how they do it at the Bonagia tonnara, I was told. I shuddered at the thought of a mattanza without voices. It would be like going to a mass where the collection plate was passed but no gospel was preached.

Although April 25 was Independence Day, a national holiday, the pace at the tonnara was picking up. Long blue boats pulled into the harbor to be loaded with floats and coils of steel cable. The loaded boats sank almost up to their gunwales. The men were getting double pay.

"What's the rush?" I asked.

"We started late this year because of bad weather," Occhiuzzi said. But Vito, in confidence, told me they were two weeks behind in the schedule because Castiglione hadn't wanted to pay the men to work on Easter and balked at renewing his lease.

That night at the Two Columns, Antonio Casablanca sat at the table next to mine with a Campari before him. He was quiet awhile, then he turned to me and said, "There's something I don't understand about you. Why do you always hang back? It makes me feel bad. Why aren't you more sociable at the Camparia?"

"I don't want to get in the way of the work," I said. "It would be terrible if the rais, even once, told me to stay away. And besides, it's my nature to be shy, at least at first."

Sicilians understand that one cannot deny one's nature. No doubt he took this knowledge back to the men; from that point on they seemed to adopt me as their mascot.

A womblike room in the back of the Two Columns was a cozy meeting place for locals in the off season, which was every season but summer. In this room the locals played cards and watched soccer matches and Formula One races on a color television on a shelf. One wall was covered with photos and posters of the men of the tonnara, pictures of the mattanza, and a giant centerfold poster of Clemente's tortured face from a French magazine. Clemente had tied coconut twine around his head; it pressed down on his brow like a crown of thorns, and he glared into the camera lens. The poster was emblazoned with the words: "Le Roi des Thons," the King of the Tunas. Below was a framed collage—torn, faded snapshots of his towheaded children when they were kids. They were a crowd.

It was in this back room that I met Vincenzo Macchì, a robust, white-haired, stocky fisherman who looked sixty but was eighty-three years old. His boisterous game of *briscola* with three young women was just breaking up as I walked in. He saw me, invited me to sit at his table, ordered me a mug of beer, and without introduction, without knowing who I was, told me the most intimate things about his life.

"I had a lover once, not a fiancée, but a lover, an Austrian girl, who came here ten years ago on vacation and fell in love with me. She loved me; I didn't love her. She was pretty. And young. Twenty-three years old. I let her move in with me. We stayed together for a year. I'd get up in the morning and go fishing, and when I came back at noon she'd have the house all straightened up and the pasta cooked. I'd sell my fish and come home and put the money on the table, *put it on the table* for her, and she would not touch it. Sometimes she'd ask me for cigarette money, and I'd give it to her.

"Era una fortuna," he said. She was a bit of luck. Vito, the tonnaroto, later told me that Ilsa would be found passed out drunk in various places all over the island, but she was never untrue to Vincenzino, a widower fifty years her senior.

It was getting close to supper time. Vincenzino walked to his daughter's *trattoria* to eat, and Antonio Casablanca, a thirty-something bachelor, went to his mother's table. Only the *aperitivo* crowd was left. Clemente hobbled in, ordered a glass of wine, and asked me what was going on at the Camparia, as if he didn't know. "No tuna in the trap yet, and everybody's on edge," I reported.

"The tuna have already passed, the big ones," Clemente said ominously.

"Why do you think that?" The tuna had been passing Favignana on the same schedule for millennia.

"That is just what I think," he said, tilting his glass and taking a sip. His arms were painted orange with Mercurochrome.

"It's still early," I said.

"The world keeps changing," he said.

One day the tonnaroti left their work to lash bouquets to the bows of the tonnara boats. The *padrino*, in a belted brown alb, walked down from the church followed by an altar boy in black robe and white smock. The boy carried a silver tankard of holy water.

Once the tonnara had its own chaplain and chapel. The tonnaroti used to pray in the Church of San Francesco, next to the Parodi palace, but now it was closed and in disrepair. Once even its priest was under the rais's command and had to be available to him twenty-four hours a day. Now the tonnara must make do with the padrino from the chiesa madre.

The priest walked around blessing the tools of the tonnara, dipping a silver wand in the vessel, which the boy proferred with both hands. The men followed him like schoolboys, hair combed with water, caps in their hands, their boats festooned with store-bought flowers. The priest sprinkled first the nets, then the carts, the coils of steel, the anchors, the ropes, the long chain, the boats at the dock, the rais, then the sea itself and the fish in it. He faced the bay and prayed God to grant the tonnaroti the strength of the bluefin, and he led them in an Our Father and a Hail Mary. Then he turned around and wished them "Buona pesca."

The men booed. They howled. They bayed in unison: "o-o-O-O-H." The priest left nonplussed. He had meant no harm, but the tonnaroti believed it was bad luck to wish fishermen a good catch. The rais chewed them out for their rudeness. The priest hadn't known about this superstition, and neither had I.

"What are you supposed to say instead?" I asked an old tonnaroto.

"It is best to say nothing," he said. "But if you must say something encouraging, you say, 'In the mouth of the wolf,' and he answers, 'May he croak.'"

The next day at 7:30 A.M. men were lined up along the dock preparing to load the last section of the Costa Alta, a mile and a half of barrier net, a marine fence that would reach a hundred feet from the sea surface to the seabed. Spread out on the asphalt work area before the Camparia gate, it covered an acre of ground.

Men sitting cross-legged on one long side attached a steel chain to the bottom. Their hands were scraped, hard, and rust-stained. The chain's weight would help keep the net from drifting and cut the need for stone anchors in half. Some fifteen men stood on the net lacing its pieces together. They seemed to be wading in a small sea. Ten men worked the top side of the net, tying it at intervals to a rope cable fitted with floats. The cool morning sun burned off the mist and became a golden disk. The work was dirty and monotonous. Five men stood in the boat and raked in the net with their hands. It took forty-three men all day to ready the net and load it. They talked and bantered over the thud of stone anchors being loaded. They spoke in their musical language, a constant twitter and chirp, like the swallows that fluttered in the eaves above them. The work was rhythmic and hypnotic. Occhiuzzi strutted around, shouting orders: "Pull! Lift it up! Stop! Go ahead! Pull, pull! The chain! The chain!" The net hissed along the pavement in fits and starts while the chain clinked the beat as it slunk onto the boat.

Inside the net house the younger men pulled more lengths of net down from the high mounds and piled them on old wooden carts with metal-rimmed wheels. One man stepped between the shafts, wrapped a tugrope over one shoulder and under the other armpit, leaned into his load, and pulled like a dray horse. The older men followed in a semicircle behind, leaning into the piled net to keep it from slipping. They dumped it in the sunshine and spread it out in long lines.

The steel cable from which the nets would hang was already stretched at sea. Along with the nets, the chain, the ropes, and the floats, the tonnaroti were also loading some of the thirty-five hundred stone anchors. Each white-gold brick weighed forty pounds. The local stone from Favignana's last working quarry was dense but too expensive, Occhiuzzi said, so Castiglione imported the stone anchors from Tràpani. They arrived in dump trucks on the ferry. For weeks two men with machetes hacked a waist into each block, then tied a cord around it.

A forklift brought the stones to the landing, where Leonardo, a young tonnaroto, bent to lift them one at a time and pushed them down a plank into the boat. Vito heaved them into the bow. Each boat held 250 blocks, five tons. Dust poured from the sandy blocks and stuck to Vito's face. Some cracked and broke as they landed. Kids with their books strapped to their backs, just let out of elementary school, had come to watch.

Occhiuzzi wouldn't sing. Once the Prima Voce sang to lighten this work. I pleaded with him, but Occhiuzzi pointed to his throat. He was hoarse from all his shouting. "You see that I cannot even speak."

The net seemed unending. It stopped only when the men attached the chain and sewed the seams together, but always more came out of the warehouse in carts. Long past the usual four-thirty quitting time the last piece of the Costa Alta was loaded. When the final bit of net snaked aboard they took off their caps.

Occhiuzzi shouted, "Forever be praised the name of Gesù!"

"Gesù!" they shouted in one voice, and the day was done.

12

The Rais

If you die a priest, a doctor, or the mayor of Favignana, no one visiting your grave a hundred years from now will know what you did in your life. But the man who has been a rais will have that word cut in his tombstone like a crown above his name.

Rais means "head" in Arabic; it has the same root as *re* and *rex*. The rais was once married to the sea the way primitive monarchs once took the land to wife. His title is no longer hereditary, but every rais is the heir of every other rais before him in an unbroken chain to prehistory.

The rais gives orders to the ciurma, the work gang that now numbers sixty-three men, oversees the construction of the trap, and decides the day to kill the tuna. The raises of Favignana have been renowned throughout the Mediterranean; they ruled the Queen of the Tonnaras.

Rais Salvatore Mercurio

On August 25, 1978, at the end of one of the worst tuna seasons ever, Rais Salvatore Mercurio died at age eighty in his home on Favignana, mourned on all three of the Egadi Islands. He was the last rais to wield complete authority; when the men formed a union he quit. Three months before his death he was interviewed about his life as a tonnaroto by the Italian journalist Gin Racheli. She published his story in her book about the islands, *Egadi: Mare e vita* (Egadi: Sea and Life). The following is her account.

The rais had retired ten years before, in 1968. He received the journalist at home, simply and affably; he had had little schooling, he spoke in deep dialect, yet he was lordly, she said. She played him a recording she'd made during a mattanza, the ciurma singing "Ai-a-mola." The voices transported the old rais once again to his musciara in that small square of turbulent sea over the Chamber of Death. He saw the singers ranged along the edge of the vascello, the long black vessel, the net hanging heavy in their hands. He repeated the imperious gesture of the raises, bending his arms at the elbow, nearly touching his shoulders, a beckoning motion, to start the pulling. Under his breath he recited the mysterious words of the song. According to Racheli, his eyes were full of tears and there was an ineffable smile on his face. Then he thanked her, saying he would never hear the song again. She said she'd come back the next year and bring a new recording. "That will be difficult," he said, and showed her to the door.

Rais Salvatore Mercurio spent fifty-seven years in the tonnara of Favignana, from 1911 to 1968. He entered when he was thirteen years old as a helper, learning the care of the nets, how to take them out of storage, how to repair them, how to store them, how to position them at sea. Later, as a full-fledged tonnaroto, he carried the heavy anchors, stones, and floats. He felt the ache of the oars in his shoulders, spread black pitch on the tonnara boats, prepared the cables, dropped the anchors, set the nets, and spent hours on his stomach counting tuna through the glass window in the bottoms of the boats. Finally came the mattanza, the ritual chants, the crazed circling of the tuna, the sprays of water, the blood, the hoarse shouts. The more numerous, big, and beautiful the tuna, the louder and more impassioned the cries of the men. When he saw the long-awaited prey, their silvery forms flashing before him, he shouted into the wind, over the sea, with the others. Before long he was covered with blood and sweat. Then the profound silence after the massacre. He knew he could never leave this work, this world.

When he was given his long gaff to grab and pull aboard the tuna, he became a full-fledged tonnaroto. Now, during the mattanza, he knew the desperate fury of the animal, felt it shake his steely arms, his legs, and his spine.

He was promoted to *arringatore*, one of the two men at the center of each group of eight men who pull up the dying tuna. The arringatori are the strongest and most experienced at killing; they use the shortest gaffs and have the most contact with the fish. They do the brunt of the lifting. Many times when he had brought a tuna to the gunwale and balanced it there he knew that the thrashing fish could break his arm or his back. He learned the classic dive forward, leaning his torso out of the boat to avoid a lethal tail swipe.

Over the years Mercurio became a great tonnaroto, strong, able to predict the movements of the fish. He was made capoguardia and continued to develop his sensitivity to and knowledge of the winds, the currents, and the water temperature. Little by little the moods and behavior of the fish held no more secrets for him. He learned to solve all kinds of problems, from breaks in the net to huge predators caught in the trap or tuna entangled in it. He took his turn at sea on night watch to warn away errant ships whose propellers might tangle in the barrier net or whose motors might frighten the tuna into breaking the trap. For fourteen hours at a time he was a human buoy in an open boat in the most inclement of weather.

The rais saw that Mercurio was a natural leader with a strong sense of discipline and duty. He offered to send him to Bengasi, in North Africa, where a new tonnara was being built. Mercurio went, and when he returned he enjoyed the respect of Vincenzo Florio, who had leased the Favignana tonnara.

In 1935 Mercurio became the rais of Favignana, commanding a hundred men. His orders were precise and unwavering. He expected to be obeyed immediately and without question. He permitted no shoddiness. He worked men hard. He might order two or three mattanzas in a day. When the men appeared tired, he sent them home to sleep at any hour of the day.

He, the rais, seemed never to sleep; nor can anyone recall that during the tuna season he spent a single night on land. He remained in his musciara, silent over the nets, conning the wind and the mood of the sea, listening for the direction of the changing currents. At night he literally felt the presence of the tuna with a fine weighted string that he held in his hand and dangled into the bastardella. The tuna brushed against the string; so many brushes, so many fish.

Time passed. The tonnara of Favignana continued to be the richest and most productive in Italy. Mercurio was famed throughout the Mediterranean for his severity and intransigence, but also for his exceptional competence. Other raises sought his advice.

In 1939 Favignana had a remarkable year: For three consecutive days the tonnaroti took 998 tuna per day. Bombing during World War II brought the catch numbers down, but in 1957, when 7,480 tuna were killed, Rais Mercurio brought the tonnara to its highest level of the decade.

The next ten years brought social change. In the sixties strange ideas crossed the strait from the Continent to disrupt the thousand-year-old hierarchical structure of the Sicilian tonnaras. Politicians, who knew nothing about the ancient trade, infiltrated first the canneries and then the ciurma, bringing new concepts of workers' rights, schedules, and adequate pay. At Favignana the troubles began with a strike in the cannery during the outstanding season of 1957.

Still, for years no worker dared question the authority of Rais Mercurio over the fishermen and the trap. But he knew that destiny was knocking, that the sacred world of the tonnara was about to change. He convinced one of his sons, a tonnaroto, to get out of the business, because that world was coming to an end. Until the end came, though, Mercurio stayed in his musciara, directing mattanzas, as long as his command was unchallenged.

In 1968 politicians tried to convince him to accept two union representatives for the ciurma. Mercurio replied that a boss is no longer a boss when intermediaries come between him and his men, and that no one except the rais could know what needed to be done, and how, and when, despite discomforts, for the good of the tonnara and so for the good of the men. He explained that capturing tuna depended on nature, not on union schedules.

But the "disease," as he called it, had already infected some of the men, and the politicians did the rest. Mercurio had said that if they named union representatives he would leave, but no one believed him because the tonnara was his life. The representatives were elected, and that same day Mercurio resigned.

It was like a king breaking his scepter. The owner begged him to stay, and so did the tonnaroti. But he had lost his direct rapport with the ciurma. His power had been limited; he felt his orders would now be scrutinized.

The old man retired to private life, and a sense of guilt fell over Favignana. Giacomo Rallo, one of Mercurio's disciples, took his place as rais. Mercurio's reputation did not fade but increased, and a mythical aura grew around him. He became a symbol of the thousands of years when tuna were a destiny and not just a business.

The newspapers covered his death, and at his funeral there were forty-five wreaths of flowers.

Rais Gioacchino Ernandes

In centuries past the rais was carried to his boat, like Montezuma or some Hawaiian king, because his feet were not to touch the ground. The crew hung garlands of flowers about his neck. The small gray musciara, the smallest boat in the fleet, was his throne. Historically, retired raises don't live long simply because they don't usually quit until they're ready to die; for the rais there is no mandatory retirement age. But when a rais retires he keeps his title, like a governor, for life.

Rais Gioacchino Ernandes, at seventy-five, kept pretty much to himself. He never set foot in the Camparia, or even near the boats, although the docks were open to everyone. But he was always discreetly on the edge of the tonnara, dressed in a tan tweed jacket and visored cap, his mustache white and trimmed. Chin up, he walked about town with a cane, slowly and bolt upright, like an abdicated king.

Gioacchino Ernandes became a tonnaroto at age sixteen in 1932, when his father was rais. Thirty-eight years later, when he was fifty-four, he became the rais, in the days when the tonnaroti would do four or five mattanzas in a year and take three or four thousand fish every time, in the days when they'd find a shark in the trap every year. "We'd let it die. The shark would drown, and we'd pull it out by hand," he said. The largest tuna he could remember during his tenure weighed 1,155 pounds.

The raises of old were extremely protective of the tonnara. "No one could visit the trap, not even the owner," Ernandes said. The men in the musciara were handpicked; they had to know what to do without orders. They had to row around the perimeter of the trap silently, no talking, no noise from shipped oars, or the wary tuna might dive deep and become uncountable.

His grandfather and father had both been raises at Favignana. The duke of Aosta came to witness one of his father's mattanzas; the nobleman stood on Florio's private boat, under an awning out of the sun. Ernandes had been rais himself for fifteen years under the Parodis.

When he was rais and the Parodis were in charge, the tonnaroti could choose for themselves the best three tuna out of every thousand taken. At the end of every operation there was a reward: a cigar for every man after they put the trap in the water; a half-liter of wine when the barrier nets were set; a little money when the Chamber of Death was ready.

Rais Ernandes quit in 1985 after just one season of working for Castiglione, who told him to move the trap farther out to sea. It had been in the same place for a hundred years, since the time of the Florios. Ernandes's grandfather, before he became rais, had been in Rais Casubolo's musciara when Florio tried the same thing. "One day they went to check the trap, and it was gone. They lost the season," Rais Ernandes said. The experiment was one of Florio's failures. The strong currents that brought the tuna had carried away the trap.

Castiglione had the same idea, to snare more fish where the currents were stronger.

"I did not accept this change of position," Rais Ernandes said. "What happened to my grandfather did not happen to me. The trap didn't disappear, but it did move. It tore apart. The tuna escaped."

"What did you do?" I asked.

"I tried to get all the material back to land. Then I left."

He was sixty-nine when he quit. A few of his men left with him, in solidarity, men who would have otherwise been asked to take his place. But the young Salvatore Spataro did not leave the ciurma, and Castiglione promoted him to rais.

The rais's authority must be unquestioned, Ernandes said. "He's the general, he's the president of the republic. If I tell you to do something, you do it even if you think it is wrong. Of course, there has to be some *tira e molla*, some give-and-take."

Now Gioacchino Ernandes sat in Piazza Europa, the piazza closest to the sea and to the Camparia, where he could watch the daily comings and goings of the tonnaroti inconspicuously. Sometimes he stepped into the one-man travel agency of his friend Giuseppe Guarassi, a tonnara buff and former director of the cannery. The old rais did his errands, made his social rounds, and sat awhile on the steps of Bar del Corso facing Florio's statue.

Rais Salvatore Spataro

In 1986 Salvatore Spataro commanded his first mattanza, the first one I ever saw. I have pictures, black-and-whites, of the young rais in jeans and a denim jacket, shouting orders from his musciara as it rocks wildly over the Chamber of Death. His black hair has turned steel gray with white streaks. It was still thick and wavy, but now started higher on his forehead. He was preoccupied and prone to angry blowups. The men called him "Il Bimbo" (the Baby) for his petulance.

Salvatore Spataro did not fraternize with the ciurma. During the day, if he was not at sea, he was usually in his windowless stone office, on the phone to Castiglione, the lights turned off, the colors drained from the incongruous posters of rainbows nailed to the bumpy, mildewed wall. Someone was always in there with him, one of his lieutenants, or the accountant from Castiglione's office in Tràpani. He fidgeted behind the desk doing paperwork, jiggled his foot, pumped his knee. The year before he quit smoking, an accomplishment that did nothing for his character, the men said.

He corrected the men in public, sometimes even in front of tourists. It was humiliating. "He should take a man aside and tell him, 'Look, you were wrong,'" said a tonnaroto who retired early because, he said, the rais showed no respect for his elders.

Last year I once saw the ciurma come back early from an operation at sea that should have lasted several hours. I watched them climb out of their boats in silence, looking hangdog, with none of the usual joking. You could have cut the air with a knife.

"What happened?" I asked Clemente.

"The rais found something wrong with the cables, and the work had to be discontinued until it is fixed. He threw his million-lire walkie-talkie into the sea and blamed Gioacchino for the mistake." I could picture the scene because I'd seen it before, men yelling at each other from their separate boats, the muscles in their necks bulging, the boat rocking with the finger-pointing, someone standing up to the rais, the rest unwilling witnesses. You can't walk away when you're in a boat. "Who knows the tonnara better than I?" Gioacchino hurled the question at the rais, making a dent in his sovereignty.

Clemente did not tell me the rais's exact words, but I knew how he was when he was angry.

"How did Gioacchino react?" I asked.

Clemente touched his finger to his eye. "He cried. Don't tell anyone."

Later Gioacchino, without mentioning the incident, told me he would rather take a punch than be chewed out at sea.

But everyone agreed the rais knew his business.

"Salvatore, you could say he was born in the Camparia," Adalgisa, his wife, told me one day over a cup of coffee in her kitchen. I went twice a week to their two-floor home in Via Mulini, a flight of steps off the Via Roma, to tutor their fourteen-year-old son, Gaspare, in English. It was a way to repay the rais for his kindness in letting me on the tonnara boats.

"His father was a tonnaroto. Salvatore was one of those who worked in the Camparia all winter, preparing the nets, the boats, for the coming spring. And he had always been in the musciara of the rais." That was how he learned, she said, watching the rais at work at close range.

Adalgisa grew up on Favignana and knew her husband from childhood. When they were engaged the Parodis discontinued the winter work to economize, so Salvatore did what so many others did: He left the island to find work. In Torino he got a production-line job at the Fiat factory, where he worked for a year and a half before marrying Adalgisa. That was twenty-one years ago. She moved north with him and started to look for a job herself.

But one month after the wedding she was one month pregnant with Diego, their first son, and she knew no one would hire her. Then her husband became ill.

"Salvatore couldn't sleep. He'd get up in the middle of the night and go walk the streets. He felt sick—terrible. There is no sea there. He went to the doctor. The doctor examined him, then said, 'There's absolutely nothing wrong with you. Go back to your *paese*.' He'd been in Torino for two years. I said, 'Let's go back to Favignana,' and he was happy. Maybe he was waiting for me to make the suggestion."

And so they came back. Salvatore found his old job in the Camparia—the Parodis had decided not to shut down in the winter after all—and he had been there ever since, all winter and every spring. But the burden of command had worn him down.

We sat in his dark office. He was looking through a notebook because I had asked him exactly how many pieces of net there were in the trap. He found the numbers and read them to me, section by section. At the end we added up the subtotals: 1,812 pieces of net.

"You can put the total in your book," he said, "but not the subtotals and the sections they belong to. This is my livelihood."

"Are you afraid that someone could use these numbers to set up a tonnara?"

He bristled.

"No. No one can do what I do," he said. "For me, this is the last year." Then he softened. "I think this is my last year. I started too young. Too much stress."

"Yes, I've watched your hair turn gray," I said.

"Ten years. Too long. Maybe over the winter I'll forget."

"You would quit as rais?" I could not imagine him standing by on this small island while another took command.

"Yes," he said. "People are always seeking me out. They want me."

"Where?"

"In Sardinia. In Libya."

"But three months away from Favignana, in the spring. . . . Could you bear it?"

"Sì. Three months. Then, for nine months. . . ." He lay his cheek in his hand, meaning he could take a nap. "In Libya they pay in dollars."

He knew the men told me all their gripes about him. They just don't understand, he said. "When I give an order, like, 'Save that piece of net, fix it,' they think I'm being stingy. Not all of them. But some of them don't understand that I try to save on materials, so if the boss asks next year, 'Can you set the tonnara without buying new materials?' I can say yes, and they will have work. Otherwise, they will be *a spasso*"—out for a walk with their hands in their pockets. "Me, no. But them, yes."

He repeated that if I printed the net subtotals he would not speak to me again. "You must not even say hello to me, in that case," he said, and got up and walked out into the light.

ꝏ

One windy day when the sea was rough and an attempt to set the Chamber of Death had been aborted, Gioacchino Cataldo kept a promise he'd made to me. The tugboat brought the single-file procession of boats back to port with an hour to spare before lunch, and the rais, in a rage, told them to all go home early. It was the day he threw his walkie-talkie away.

Gioacchino walked off the dock and straight to me. "Today would be a good day to go to the cemetery," he said.

He always visited on the Feast of the Dead, November 2, but because I had asked we went that spring day. He swung his leg over his bicycle and told me to follow him on mine. I waited for him in the piazza while he bought flowers. He came out of the florist shop with a thick bouquet of red carnations and pedaled ahead of me down the narrow cobblestone street that leads to the Lungomare, the road along the sea. He motioned for me to coast alongside him and pulled a lemon from his pocket, stuck his thumbnail in it, and handed it to me as we rode. "Smell it," he said. "It's like a lemon from two thousand years ago. No chemicals, grown on this island in the salt air."

In ten minutes we were at the cemetery. Its walls are ten feet high, made of island sandstone. It has two iron grille gates, front and back. The rear gate frames the sea and a quiet rocky cove. He told me to lock my bike. I asked about the aborted operation at sea, but he hushed me. "This is not the place," he said.

"When I was a kid, I used to be afraid of the cemetery," he said. "Now I don't fear the dead, I fear the living."

The weeds had grown high around the graves. Gioacchino complained to Rosa, the caretaker.

"Cutting weeds is man's work," she said.

He knew his way around the cemetery; to him it was a neighborhood, the home of old friends he visited often. I followed Gioacchino around the graveyard.

First he took me to the tomb of a sailor, a navy man, who had saved his ship when it was in flames, at the cost of his own life. Gioacchino laid a carnation on his grave. Then he walked me over to his father's grave, emptied the dead flowers from a vase, and replaced them with fresh carnations. He took me to see the gravestone of a man who had lived a hundred years on Favignana and died on the day of his birth. Fixed to each gravestone was a formal black-and-white portrait, oval under glass, of the person buried below.

We bushwhacked through the weeds and wildflowers to visit the unimposing tombstones of the five deceased raises of this century. They were scattered throughout the cemetery, each in his own family plot. To Gioacchino they were all family, whether by blood or in spirit. He walked around and laid a red carnation on each of their graves: Giacomo Rallo, his own godfather; Salvatore Mercurio; Flaminio Ernandes; Gioacchino Ernandes, Flaminio's father; and Gaspare Grimaudo, rais from 1900 to 1907, his grave untended, his people all gone.

Grimaudo's face was thin and bony, almost skeletal, as if the photograph had been taken in the nick of time. But the word RAIS was carved deep in capital letters in a semicircle above his name, like a crown.

His gravestone leaned hard to the left, displaced by the roots of two splendid wild caper bushes, thigh-high and in full bloom. But they were toppling the rais's stone. Gioacchino bent down, and with one huge hand he grabbed their stems near the ground and pulled them up, roots and all.

13

The Trap

Cristina and Rino had remodeled the Two Columns over the winter, and now Cristina, blonde and petite, was redecorating before the busy summer. She stood on a bench hammering nails into the arched tufa entrance to the back room. Then she draped an old coconut fiber net over the nails. Next to the bar hung a framed black-and-white print of her father and Filippo Messina in the 1950s standing barefoot in a vascello, loaded to the gunwales with giant bluefins they had just killed. And there was another boat behind them, also mounded with tuna. They clutched their gaffs and grinned. They were young and still had all their hair and teeth. Messina held out his hand as if to ask the photographer for payment. The frizzy gold cords of the net brushed the frame.

Clemente was at the bar with his aperitif, picking at his scabs. He wore a Birkenstock sandal on one foot and a flip-flop on the one that was still swollen from the motorcycle accident.

"Do you want me to steady you?" he asked Cristina. He put his square hand on her hip, turned around, and grinned at me.

"Sex maniac," Cristina said. They were old friends. Cristina adored him. Clemente was strength and kindness. He harbored a deep unspoken love for his island. He trusted few people; Cristina was one of them.

"Who gave you that net?" he asked, knowing it came from some hidden stash at the Camparia, most likely his own.

"The Parodis gave it to me."

Clemente's company was in great demand. He was shy, but onstage at the Four Roses dance hall he would recite, sing, and dress in drag. All the regular off-islanders who rented houses on Favignana wanted Clemente to dive with them, Clemente in their boats all day, Clemente at their café tables, their parties, and their fish roasts. Cristina never charged Clemente for his wine; island icons don't pay for their drink.

I remembered seeing him once, before I knew him, bicycling right into the Two Columns with a baby girl in his handlebar basket. She had a pink ribbon tied around her fuzzy blonde head. Clemente was smiling, his eyes twinkled. Cristina told me that the baby was his daughter, Callista. Her mother was German.

Now Clemente hobbled to a table and sat down. Ninetta, the twenty-something beauty who managed the yacht harbor in the summer, finished her coffee and walked over to him. "Ciao," she said. Lifting a lock from his forehead she planted a tender kiss on his right temple. He took it stoically.

A white mist veiled the sky. We were both glum that day. The soft light was a relief from the hard glare of the Sicilian sun, but on Favignana a spring day without sunshine is depressing. The rais was dropping anchors, but neither of us could work: Clemente was still limping, and I was not allowed to go to sea. The boats were crammed with giant anchors, cables, and rocks, the men had to concentrate, and it was dangerous, so the rais would not let me go with them.

The rais had already stretched the steel cable from which La Coda would hang. Today he would secure that cable with pairs of anchors along its length. I knew that where they were working today was close to shore.

I said, "Clemente, take me to the cemetery so we can watch *u cruciatu*."

He stood up and limped out to his tiny white Fiat 500, which barely contained him. As we left the piazza and headed toward the sea, I noticed people staring at us and mentioned it.

"They are staring at me," Clemente explained. Perhaps he just wanted to put me at ease. "People always stare, wherever I go. Why is that?" he asked. "Once I went into a restaurant in Munich, and people at a table on the other side of the room stood up to get a better look at me."

I studied him a moment. It wasn't only his hair, but the shock of opposites: his rocklike body with now a rock-hard gut and those absurd thick curls foaming about his broad, square face, the lined, worried face of a middle-aged man who was once transcendentally handsome. I'd seen the old pictures, so I knew.

"Why is it? Why do people stare?" he asked.

"It's your powerful charisma," I judged. I always fall for charismatic types. This was dangerous, I thought. I shouldn't do it. But then I remembered a headline, "Maggio Goes with the Flow," that appeared on the front page of my hometown paper after I had sent back an eyewitness account of an eruption on Mount Etna. In an instant my course was set.

"What is this 'charisma'?" he asked.

"It comes from a Greek word meaning 'gift,' like a gift from the gods. It's a spiritual thing, a kind of energy that surrounds you and draws people to you. John F. Kennedy had charisma."

He seemed pleased.

We drove past the town's walled cemetery to the Phoenician oven tombs and parked there. The sea was choppy, and small waves broke in frilly lines. On the horizon we could see the boats performing *u cruciatu*.

It is a ballet at sea, with a pair of swan boats circling opposite each other in synchrony, the anchors dropping like molted feathers. The rais stands between them in his gray musciara as his crew pull his boat along under the cable. They lift the cable floats up over the bow, then off the stern back into the sea. The anchor boats, pulled by the towboats, drop pairs of anchors every fifteen meters along the cable's length.

All eyes were on the rais, who raised his arms above his head as the boats circled out and away from him. When the anchor boats were fifty yards out on either side, he dropped his arms and each boat dropped an anchor.

Then they circled back in, fifteen meters farther up the cable, attached fifty-yard ropes to two more anchors, and threw the anchor ropes to the musciara. The men in the rais's boat attached these to the cable. The process was repeated all day.

I watched through my telephoto lens, sometimes passing my camera to Clemente. I had the quintessential tonnaroto sitting next to me with no one barking orders at him, no distractions. I took out my notebook and asked Clemente to draw a plan of the trap. He sketched the many-roomed cage, enormous and intricate, with the two barrier nets funneling the migrating tuna into it.

"Eight years ago," Clemente said as he drew, "a big oil tanker carrying the Panamanian flag took La Costa to Marsala." The ship was huge, he said, six thousand tons. Either the captain didn't see the yellow warning boat or he didn't receive the warning from the night watch crew. The net caught in the ship's propeller, which ripped up hundreds of tons of anchors, cables, and nets. Insurance paid for the damage, Clemente said, "but it took us three days to replace this bit." The sketch in my textbook became an important reference document. I asked him to sign it. He did.

"You all seem to dislike Castiglione," I said to Clemente, "but he must love the tonnara too if he keeps it going in the traditional way. He can't make much money at it, considering his expenses. And he has replaced the old boats and the nets. He's bought plastic floats, added new cable and chains. Even though he cut the workforce, he must have done it just to stay alive in this business."

"Don't kid yourself, he's in it for the money," he said. "Figure 1,000 tuna, an average weight of 150 kilos per tuna. That's 150,000 kilos of tuna in a year. The Japanese buy it from him at about 12,000 lire per kilo." That worked out to 1.8 trillion lire of gross profit, about $1.25 million. "Subtract from that 600 million lire in expenses"—for the the lease, salaries, materials, and insurance—"and what do you get?" I scribbled and figured about $733,000 clear. That did not include the government subsidy money Castligione was supposed to be getting.

"If the tonnaras are so profitable, why did Formica close, and Scopello, and all the other tonnaras in Sicily, except Bonagia?" I asked.

"Because Favignana is the farthest out to sea. It is fortunately situated, because the biggest shoals of tuna pass here," Clemente said.

The waters around Clemente's island have just the right salinity and temperature. Washed by strong currents, they are limpid and clean. Because of its location, Favignana's tonnara survived on the last trickle of what was once a great river of tuna.

Clemente said he made only about $5,000 during this three-month period of the tonnara, and that was with the extra $1.50 a day he earned for being a boat captain.

"That's good money, but not enough to last a year with a family," Clemente said. He lived with his two youngest sons, now men themselves, in an apartment a half-mile from the piazza. And he had daughters who would want weddings. The tonnaroti were paid thirty-five dollars a day for the three months, plus a bonus of about thirty cents for every fish the crew pulled aboard, including swordfish and tuna that died prematurely, tangled in the net. That was about three hundred dollars more for every man when a thousand fish were caught.

That evening at the Two Columns I was going over my notes when Occhiuzzi walked in. He stood behind me and looked over my shoulder at what I'd written, scribbled half in English, half in Italian. "Stop!" he said when I turned a page and he saw Clemente's sketch of the tuna trap.

"That's wrong. That is all wrong," he said. "Who drew this for you? Clemente?"

"I asked him only for a rough sketch. This is not a definitive drawing, I understand that."

Occhiuzzi said he had been at sea all day. "Only you and Clemente were missing today," he said. He ridiculed the drawing in front of Antonio Casablanca. He wanted to rip the page from my notebook, but I wouldn't let him, even though he promised to bring it back. "Ha! Fifteen years a tonnaroto, and he doesn't understand a thing!" Occhiuzzi said. "Tomorrow they will load another boat with anchors and nets, and the next day they will lower them into the sea. I'll put in a good word for you with the rais."

Clemente brought supper to my house that night—a thick steak, some lettuce and tomatoes, half a loaf of round bread. He arrived in a Hawaiian shirt, smelling of aftershave. I was a vegetarian, so I put some water on for pasta. When it boiled he stuck his finger into the pot to taste it. He added salt, broke up a pound of thick buccatini with his hands—the Sicilian equivalent of ripping the Manhattan phone book in half—glugged oil into a pan, cut up garlic, parsley, and peppers, and started to fry them. When they were done, we ate and I tested the waters.

"I saw you once with a baby in your bicycle basket. Who was she?"

"Callista, my daughter. Her mother is a German lady."

"What's her name?"

"Margo."

"What happened to Margo?"

"We lived together for several years, then we broke up and she took the baby home with her to Munich." When he met Margo he'd been living with another woman for ten years. "You remind me of her," he said. He missed the child, he said. He wanted Callista to speak Italian. But he swore he didn't love Margo anymore.

They had fallen in love the first time he saw Margo on the beach. "She was a doll," twenty years younger than him. She could not speak Italian. Clemente began to teach her.

When someone told his live-in girlfriend he had been seen with a pretty German tourist, she packed up and left abruptly. Though she had helped Clemente raise his brood of children, he never heard from her again. "I still miss her," he said.

Margo left her job in a German travel agency and moved in with Clemente. He taught her how to crew on his private boat, the *Poseidon*, and in the summer they took tourists on joyrides. "She was a better sailor than I was," he said. Then the baby came, and they began to fight. I didn't ask for details. They broke up, but they were friends now. In the winter he goes to Germany to visit. "In the spring she takes a vacation and brings the baby down here around my birthday." Clemente, like Gioacchino, was born at the height of tuna season.

I had no corkscrew for the wine he'd brought, so he asked me for a towel and folded it into quarters. He braced his arm against the kitchen wall, stood spread-eagled for balance, and pounded the bottle bottom against the towel on the wall, repeatedly, checking after each blow to see if the cork had loosened. The process took five minutes, but eventually the gas pressure pushed the cork out of the bottle neck. It was like watching a caveman open a bottle of Corvo.

"Did you think we would die of thirst in a desert because we had no corkscrew?" he asked. Clemente knew why I had invited him. My inquisition must have confirmed it. He was free, and so was I. I took him and two glasses to my square stone room.

I placed my notebook on my bedside table and interviewed him in bed.

A man like Clemente was something that only Favignana could produce. For example, he had a five-inch scar on his shoulder in the shape of a bluefin. "When I was seven I was playing in the sand on Lido Burrone and fell on a broken bottle." He used to spear fish with the spokes of a broken umbrella frame. I took his spatulate hands in mine and turned them over. The lines on his palms—head, heart, life, and fate—were etched deep and crisp, with no feathery fringes, which most people have. I lay my head on his chest; his heart rang like a bell.

Clemente got his first job when he was six, driving an ox in circles around a well that irrigated fields, eight hours a day. He said his mother was so strong she could deck a man. His father, a quarryman, had gone away when Clemente was ten. The oldest of five siblings,

Clemente became the head of the family and went to work at the cannery.

"I cut, I cleaned, I carried, I took out the spine. I spent fifteen, twenty years in the factory," he said. When it closed he was thirty-seven.

"That's when I became a tonnaroto. It was the sea that drew me to this work. It was work I liked," he said. He was strong and unafraid, so they made him arringatore the first time he killed. "It's a battle between you and the animal. If you make a mistake, you pay," he said.

He married young. "At nineteen I already had a son," he said. Then six more children in quick succession. Separated from his wife, he raised the children himself, he said, except for the youngest, a daughter, whom his cousin adopted as an infant. Now she was a beautiful teenager. Women fell hard for Clemente, and over the years he had fathered two more children by two different women.

"How old were you the first time you made love?"

"Fifteen," he said.

"Where did you do it?"

"By the sea, at San Giuseppe," near the turn where his bike skidded off the road. I ran my finger down his arm. His cuts were healing. I found one tiny blue tattoo, A+, his blood type, on his bicep. Very practical. He was familiar with bloodshed.

Once, during a mattanza, Clemente slipped on tuna blood and fell into the Chamber of Death, roiling with 350 bluefin. "I landed astride a big one," he said. "I was not as afraid of the fish as I was of my colleagues," who proferred him the barbed ends of their gaffs to grab on to.

Instead he made his way back to the vascello and climbed up Il Coppo's thick weave like a ladder.

Clemente was a man who could look a giant in the eye, pull it from the sea, and hold it thrashing, panicked, and fighting for its life on the end of a barbed wooden stave. What did it feel like to kill a giant bluefin with your bare hands? I could never know; I could only ask, over and over. This time it was my own fault the pillow talk ran to fish.

"What impressed you more? The first time you made love or the first time you killed a tuna?"

"Making love." We got between the sheets.

"Clemente," I said.

"Sì?"

"Let's say it's the morning of the mattanza. You pick up your gaff in the Camparia and walk toward your boat. How do you feel? What's going through your mind?"

"Well, the first mattanza of the year, you're scared. There's a lot of tension. Because if you make a mistake, you could get killed by a *codata*, the blow of the tuna's tail. After the first one you're not scared anymore. However, it's not something you do every day."

"Okay," I said. "Now you're beside the Chamber of Death. The net is pulled up tight, the fish have all risen to the surface, the thrashing has died down, and the rais gives the order to kill. How do you feel, looking the fish in the eye? Do you feel any sympathy for the fish?"

"You do it because it's survival. You do it to live. Or you don't choose this life. You become a banker. It's not for the violence. It's not something I do for pleasure, or to please others. It's survival."

We were a secret unit for a matter of weeks. But when Margo came with the baby for ten days I saw no more of him at my house. She was the mother of his child and deserved a certain amount of respect. I assumed he was sleeping with her. She brought the baby in a stroller to the Camparia during the day. Callista was a toddler now, with Clemente's blond curls and squat physique.

Although I arrived in time to see the barrier nets set, I was too late to see the snare itself go into the sea. But I'd read about it.

The rais founds the trap the way Alexander the Great founded cities. Alexander drew the boundary lines of Alexandria with cornmeal: Here will stand the city walls; here a temple; here the theater. The rais traces the boundaries of his springtime realm with two thick coconut cords stretched along the boundaries of the trap. This act, reserved to the rais, is called *l'incrociare*.

On a day in April he is towed three kilometers out to sea and chooses the site using the wisdom of the raises before him and his own knowledge of the sea bottom, the currents, the habits and highways of the tuna. Too far out and the trap may be damaged or displaced by strong currents; too close to land and the tuna may not enter.

Piero Corrao, the fisherman who first brought me to Favignana, knew where to drop his nets by lining up Mondello's mountain peaks in a certain way, a skill he learned from his father when he was just thirteen. He knew the best spots on the bay for catching different species and positioned his boat on the unmarked sea with an accuracy that amazed me. Once he told me about a man who had dropped his gold watch when he removed his sweater aboard Piero's boat, the *Francesca*. The next day Piero returned to the same spot with a scuba diver, tied a rock to a rope, and dropped it overboard. The diver followed the line down and found the watch next to the rock. Piero was able to return it to its owner.

The trap is unseen architecture; it is whole only when submerged. To help me picture it, I spent hours in the Favignana library poring over old books and artists' renderings. But they were pictures of traps centuries old, or of traps other than Favignana's. One day I showed the rais a drawing I'd found.

"No, no. That's all wrong," he said. To my surprise, he took the pen from my hand and began to draw his trap in my notebook. The two pages together weren't wide enough, so he found some longer paper and traced it out.

The trap is oblong except for a widening at the shoulders that makes it look like a wooden coffin. It is divided into seven rooms by net walls with gates in them. The easternmost room is called Levante. It is fifty meters squared. One of the palms I saw at my first mattanza stands above its easternmost point, so that even from land the rais can see if the trap has been displaced by the currents. Once, when the bluefin were plentiful, the tonnaroti used this room to hold the first shoals that entered.

The next room is the largest, the *Camera Grande*, 110 meters long and 64 meters wide, the widest point of the trap. Tuna enter this Great Room through the *bocca di nassa*, the trap's only entrance. Two net panels form a V-shaped funnel with its apex inside the Great Room, making it difficult for the tuna to turn around and swim out.

The *bordonaio* is west of the Camera Grande, then the *bastardo*, 50 meters long, then the *camera*, also 50 meters long, then the *bastardella*, just 25 meters long, the last room before the *Camera della Morte*. The Chamber of Death is 100 meters long but just 22.5 meters wide, the length of the tonnara's longest vessel. It is the only room with a net floor. On the morning of the mattanza the fish are chased from the bastardella through a flowered gate into the Chamber of Death.

It is possible but unlikely that tuna will enter the trap before the barrier nets are in place. The net La Costa stretches from a point near the island's cemetery out toward the small, uninhabited island of Maraone. It consists of three parts: the Costa Bassa, 1,500 meters long; the Costa Alta, 2,000 meters long; and the Terza Costa, known on Favignana as La Spada di Orlando (Orlando's Sword), 1,500 meters long.

The second barrier net is La Coda, the tail. It runs from the west side of the trap entrance to the port on the north coast of Favignana, three kilometers in all. All along both barrier nets are constructed *rrioti*. If these could be seen from the air, they would look like giant, squared-off fish hooks. A *rrioto* steers the milling shoal toward the trap entrance. The island itself acts as a barrier leading the fish to the bocca di nassa.

The trap is a sacred space that requires silence and reverence. In the old days the crew of the rais's musciara was chosen for the ability to act without orders, for there was to be no talking during *l'agguaetare*, the act of counting fish.

Because it is so tightly woven, Il Coppo, the floor net of the Chamber of Death, can catch the currents like a sail and rip the trap apart at its seams, so it is installed at the last possible moment in the season, now just a few days before the first mattanza. In the old days the tonnaroti would set Il Coppo the very morning of the first mattanza. It is left in place for the rest of the season, but the rais worries about it.

There were miles more of barrier net to go up, but the rais was adamant that I could not go to sea with the men when they dropped anchors. Occhiuzzi just shrugged and held up his palms when he saw me looking forlornly at the loaded boats about to leave port.

I had offered to buy a few cases of cold beer for the men to enjoy at the end of the long day, but I had to get the rais's permission to do so. The rais had to approve anything that concerned the tonnaroti. "I will not allow it," he had told me. "You should just have fun."

The rais was more agitated than usual now. I wondered if he believed Clemente's pronouncement, that the big fish had already passed. The trap was still far from complete, and this made him feel a little incomplete himself. There was still much work to be done before they could set up *I Santi*.

I'd seen the cross up close for the first time this year. It lay patiently on blocks in the black shadows of the boathouse. Ten feet tall, it bore colored pictures of the saints who protected the Favignana tonnara, and a bronze statue of Saint Peter on the papal throne. A plume of fresh palm fronds, newly blessed on Holy Thursday, sprouted from the top of the cross, and a bouquet of white calla lilies and gladioli was lashed behind the palms.

The rais was out of sight, being towed to the trap. At the dock three longboats were loaded with anchors, chains, and cables. Clemente walked close by me and said, in a low voice, "Are you going to hire a boat to see this operation?"

I hadn't thought of that. I hotfooted it over to the fishermen's port near the ferry landing and asked the first man I saw how to hire a boat. Sebastiano Massiguerra, eighty, offered his services. He wore a white turtleneck and sweater, and he'd just come in from a morning's fishing in his ten-foot boat. He charged twenty dollars. I used the beer money to pay for it. He thought I was a tourist who wanted to tour the island's marine grottoes, but when I explained I wanted to see *u cruciatu* at sea, he put me in the bow, manned the outboard motor, and told me to watch for cables.

When I saw one he drew in the motor and lifted the rudder, and we coasted across it. We approached the first anchor boat. I panicked when I saw Matteo Campo through my telephoto lens, waving us off. "Don't worry, he was just joking," Zu Sebastiano said.

He'd been a tonnaroto himself thirty years before, so I trusted his judgment. I watched in fascination as the men pushed the huge anchors overboard, the lines slipping into the deep after them, but Zu Sebastiano had seen it all before and decided to fish. He dumped tiny white snails from a glass jar and bashed their shells with a rock to get at their slimy slug bodies. Then he baited a hook. We stayed out two hours.

"You know, it's friendship that's important, not money. Every year there are eight Romans who come down to the island for their vacation. They make barbecues on the beach, they never eat in restaurants, and they invite me to their suppers. That's living," he said.

He accepted my money when we got back to port. Then he thanked me for a fun afternoon and kissed me on both cheeks.

My fears about not being accepted by the men had proved to be unfounded. By the week after my arrival they were already trying to run my life. One morning I went to the floating dock to photograph the rais's musciara as it sat empty on the calm water in the golden pink morning sun. I thought I was alone until Zu Massino popped up into my frame, rag in hand. He'd been cleaning the glass through which the rais and his men counted fish.

"Why don't you marry?" he said, apropos of nothing. I'd come up against this before in Sicily.

"Because I like to be independent," I said. I gave him an argument I thought invincible in Sicily: "If I had a husband and children, how could I abandon them and come here for six weeks?"

"It can be done," he said. "You make yourself suffer your whole life to be free for six weeks?"

"Not just these six weeks. I travel often."

"Still, you should marry." He stared right through me for a minute. "Even priests should marry, make children for the Church." This was a radical thought.

I walked back to the Camparia to hear the latest. A thirteen-hundred-pound fish had been found tangled in the nets that morning. It was a good sign. It gave the men hope, and the rais's tension seemed to melt a little. Giuseppe Aiello, the gateman, said, "Let's hope it's a *buon' annata*." A good year was a year of many tuna.

14

The Prey

Marine biologists marvel at the bluefin, built for speed and stamina. "A description of migratory patterns of fish would not be complete unless the remarkable feats and movements of the bluefin tuna, *thunnus thynnus*, were included," said Brian A. McKeown in *Fish Migration*. The magnificent fish now swimming toward the trap are up to twelve feet long, with silvery blue sides mottled with yellow and splashed with violet. Their dorsal finlets run down their spines in a line of perfect yellow triangles.

A robot bluefin designed by a student at the Massachusetts Institute of Technology has been studied for the design of its upright, lunatc tail, whose engineering may one day replace rotary propellers on submarines. On either side of the tuna's caudal fin is a keel that creates turbulence, which lowers the drag on the tail fin, a model for the design of torpedoes.

But bluefin sprint faster than torpedoes. The largest are as big as sports cars and can accelerate from zero to sixty in ten seconds—in water, a medium eight hundred times denser than air. Bluefin are the second-largest bony fish in the ocean, after the black marlin, which reach a mature weight of two thousand pounds. When the bluefin hit top speed, their pelvic, pectoral, and front dorsal fins all retract into slots. They use ram ventilation, as do jet engines, swimming with their mouths open to get oxygen. Like sharks, they suffocate if they swim too slowly.

Bluefin grow quickly and can live twenty to thirty years. An eight-year-old typically weighs 270 pounds. Giants are at least 76 inches long and can weigh three-quarters of a ton. One weighing 1,496 pounds, aged thirty-two, has been recorded.

The ancient Greeks called the bluefin the sea pig and thought that it grew so large by eating acorns dropped from submarine oak trees. Adult bluefin actually eat mackerel, sardines, squid, and anchovies, and the largest of them also eat cod, eels, and cephalopods. But, often to their misfortune, the bluefin will snap at anything moving. Biologists who studied the stomach contents of a 150-pound bluefin caught eighty miles off the New Jersey coast found elastic ponytail holders, two cocaine inhalation straws, monofilament fishing line, fragments of drinking straws, pieces of balloons and Ziploc bags, parts of pens and markers, and bands used to bind newspapers for delivery.

Research suggests that bluefin hunt cooperatively, like wolves or lions. They swim in surface waters half the time. The first investigation of the structure of schools of predatory fish was carried out by Brian Partridge, Jonas Johansson, and John Kalish, who analyzed aerial photographs of Atlantic bluefin in the wild. The parabolic shape of schools suggested team hunting, with individuals positioning themselves within the schools to catch whatever the tuna ahead of them had missed. This formation also cut down on drag.

Bluefin schools take many shapes. Of the 141 schools photographed, 26 took the form of a parabola, 23 formed shoulder-to-shoulder echelons, and 16 swam in head-to-tail lines. Bluefin exhibit the most rigidly defined school structure yet observed. Scientists have found a system of lateral lines in bluefin skin that is exquisitely sensitive to changes in water pressure, and these may account for their ability to swim at speed in intricate and ever-changing school formations.

Douglas Whynott described bluefin schools in his 1995 book *Giant Bluefin*. While aboard a commercial tuna boat off the coast of New England, he saw school formations change from minute to minute. One school of two hundred was milling in a cartwheel. The lead fish broke off the edge, followed by another, then another, until an echelon formed. Then there was "a great diagonal line pulling away from the mill like a thread coming off a spool."

Bluefin are warm-blooded, with a very muscular heart, large blood volume, and high hemoglobin concentration.

Because they can maintain body heat, the bluefin can swim in colder waters to expand their hunting territories. They have small teeth, good hearing, and binocular vision through huge eyes that never close. Unlike most fish, the bluefin's eyes are flush with their bodies, another drag-reducing feature. They can deposit fat and use the energy for their epic journeys. They migrate a hundred miles a day through ecosystems ranging from tropical to sub-Arctic.

They are superlative animals; their only predators are sharks, killer whales, and humans.

The bluefin can cross an ocean basin in a matter of weeks or months, making it a difficult fish to study. In 325 B.C. Aristotle theorized that bluefin were an Atlantic fish that spawned in the Black Sea. Most modern experts say bluefin spawn in just two places—the Gulf of Mexico and the Mediterranean. Those that breed in the Gulf of Mexico are generally believed to roam the western Atlantic, while those that spawn in the Mediterranean are said to roam the eastern Atlantic. The boundary between the regions is forty-five degrees west meridian. As a result of this prevailing two-stock theory, the International Commission for the Conservation of Atlantic Tuna (ICCAT) has managed the eastern and western bluefin as separate fisheries.

As of 1998, ICCAT's catch quota for the western stock—bluefin caught primarily by U.S. and Canadian fishermen—was set at 2,394 metric tons, and the eastern stock quota at 40,494 metric tons. The Europeans have less stringent fishing restrictions because the eastern stock is seen as less endangered.

But that view is now under scrutiny by policymakers and scientists. A 1994 National Research Council review called for studies of the extent of bluefin movement between the western and eastern regions. Recent research has begun to test the two-stock theory; if the rate of exchange between the two populations is high, management strategies may have to change to protect the species.

In independent studies over the past forty years the researchers Frank Mather III of the Woods Hole Oceanographic Institution on Cape Cod and Raimondo Sarà, a Sicilian marine biologist, agree that there is a 5 percent exchange between the two stocks. Mather and Sarà used traditional tags and depended on cooperative fishermen to return the markers. But recently some biologists have come to suspect that just one stock roves all of the Atlantic.

Using microprocessor-controlled archival tags powered by lithium batteries, the latest tool for wildlife tracking, the Stanford University biologist Barbara A. Block and her colleagues produced the first moment-by-moment chronicle of six months in a bluefin's migration. Block surgically implanted the devices in thirty-seven bluefin caught with rod and reel off Cape Hatteras, North Carolina, in the winters of 1996 and 1997, then released the fish again into the wild. The tags, which recorded the bluefin's locations, dive depths, body temperatures, and ambient water temperature every two minutes, were programmed to pop off the fish and float to the surface at designated times. The devices transmitted these data to a satellite that, in turn, relayed the information to Block as e-mail. One bluefin dove more than twenty-four hundred feet and maintained an eighty-degree body temperature in forty-degree water for two hours.

More important to policymakers, the project gave the first hard evidence of the initial migration path taken by western Atlantic bluefin as they fanned out into the Gulf Stream. Within ninety days two tagged bluefin crossed between management zones, and four more were within five degrees longitude of the stock boundary meridian, Block reported. "These results confirm that Atlantic bluefin from the 'western' stock are subject to eastern fishing pressure, raising the question of how often such cross-boundary movements occur among mature individuals," Block wrote. If further study shows that the mixing rate is high, all bluefin-fishing nations may need to heed the same strict quotas to ensure the animal's survival.

The spawning stock of the Atlantic bluefin is now only 10 to 30 percent of what it was in 1970, according to the National Marine Fisheries Service. Biologist Carl Safina reports that the west Atlantic breeding population has dropped 90 percent since 1975—from some 250,000 animals to about 22,000—and that the east Atlantic adult population is now estimated to be half what it was in 1970.

Sport fishing for bluefin first became popular in Europe and on the East Coast of the United States in the 1920s. In February 1922, the *Toronto Star Weekly* published a dispatch from their roving European reporter, Ernest Hemingway, under the headline "At Vigo, in Spain, Is Where You Catch the Silver and Blue Tuna, the King of All Fish." Hemingway wrote: "The Spanish boatmen will take you out to fish for them for a dollar a day. There are plenty of tuna and they take the bait. It is a back-sickening, sinew-straining, man-sized job even with a rod that looks like a hoe handle. But if you land a big tuna after a six-hour fight, fight him man against fish when your muscles are nauseated with the unceasing strain, and finally bring him up alongside the boat, green-blue and silver in the lazy ocean, you will be purified and be able to enter unabashed into the presence of the very elder gods and they will make you welcome."

In the 1950s and 1960s the area around Martha's Vineyard was famous for tuna weighing between one-half and three-quarters of a ton. The giants—then called horse mackerel—made for spectacular dockside trophy photos. After the snapshot the bluefin were sold for pet food at pennies per pound.

But sport fishing was not responsible for the huge decline in stock. Commercial fishing was. By the mid-1970s the popularity of bluefin as a food item had reached epic proportions.

Technical innovations fueled by increasing demand in the 1970s were responsible for the decline of the world's fish populations. Fishermen from all nations could fish farther from shore, using radar to locate schools. Satellite navigation systems helped them return over and over again to prime spots. Factory ships could deploy nets so huge they could swallow twelve jumbo jets in a single gulp, according to *Time* magazine. These purse seines and long lines have perhaps done the most to ensure the bluefin's demise.

A purse seine is a cylindrical net suspended from surface floats and dropped over a school. Having corralled the tuna, the ship then winches in a cable to close the bottom of the net like a purse drawstring, trapping hundreds of fish in seconds. Long lines consist of a primary line twenty-five to fifty miles long supporting secondary lines containing hundreds of baited hooks. Before seine fishing became widespread in the 1970s only 5 percent of researchers' tags were returned. At the height of seine fishing 48 percent of the tags came back—nearly half the tagged bluefin were being caught.

Frank Mather, whose home was filled with pictures of him and his wife game fishing, said that the bluefin is so far-ranging that "it would be difficult to catch the last one." The tuna is not at the point of biological extinction, Mather said, but economic extinction is near.

Not everyone agrees. On the U.S. coast, where the bluefin is hunted with rods and reels, spotter planes, electrified harpoons, and purse seine nets, accurate population figures are hard to come by. In *Giant Bluefin*, Douglas Whynott writes of two spotter pilots who, on August 8, 1993, displayed aerial photographs of bluefin swimming on the surface in the Gulf of Maine. "Excluding those with any hint of repetition, the two pilots' photos from that day showed 4,984 fish," or "a quarter of the estimated stock of giant bluefin in the entire western Atlantic."

Yet already in the 1960s concern over declining stocks led to the formation of ICCAT, which assumed scientific and management authority for tuna and other pelagic species such as the marlin and swordfish. ICCAT's stated mission is to manage the bluefin on the basis of maximum sustainable yield.

The scientist Carl Safina, author of *Song for the Blue Ocean*, himself a former ICCAT delegate, is an outspoken critic of the commission, which he claims has been too cozy with the fishing industry. For example, one of ICCAT's U.S. commissioners was also employed full-time as a lobbyist for the national seafood industry, he wrote, and another commissioner worked for the U.S. Department of Commerce. Safina has said ICCAT has a history of setting quotas at levels above accepted maximum sustainable yields, ignoring even the advice of its own scientists, who calculated that ICCAT's bluefin quota presented a 10 percent risk of extinction within ten years.

"Until the Commission becomes serious about complying with its charter mandate to manage for sustainable yield, the acronym ICCAT will appear to represent International Commission to Catch All the Tuna," Safina wrote in *Conservation Biology*.

Safina has also questioned the commission's effectiveness: Nonmember countries may catch more than 80 percent of what member countries take in the Atlantic, according to Japan Fisheries Agency officials, and some tuna boats of member countries have been reported flying flags of convenience of nonmember countries to avoid restrictions, Safina said.

Raimondo Sarà has studied the eastern Atlantic bluefin through close association with the raises of the Mediterranean tonnaras for fifty years. He said the tuna arriving in the Mediterranean to spawn swim in surface layers, following branches of Atlantic currents or countercurrents, which usually follow established routes. The displacement of the currents by winds—toward or away from the traps—has a critical effect on the catches. West-flowing currents bring the tuna to Favignana's trap from the east. The tonnaroti of Favignana said the fish swim *occhio a sinistra*—with their left eye toward land.

In the spring the bluefin swim through the Strait of Gibraltar, females in the lead, pushed by the urge to procreate. The females are each carrying from one million to thirty million eggs, depending on their size.

The tuna speed toward the trap, schooling, mating, and dodging sharks. No one is sure what they use for a compass; they may steer by memory, by visual and chemical cues, by anomalies in Earth's magnetic field, by temperature differences, by salinity gradients, by the strength and direction of the sea currents, or by changes in the electrical field and hydrostatic pressure differences.

Scientists do not know how the bluefin find their way back to the coast of Favignana, and the tonnaroti do not ask why they return every spring. They only wait for their trap to fill.

I Santi

When the trap is set, the saints take the watch.

One morning the tonnaroti swung open the iron grille gate at the back of the boathouse. Clemente, now working again, rolled his yellow sweatpants up to his knees and waded into the icy spring sea to help his crew pull the boat named *Marte*, after the planet Mars, out of dry dock and into the water. Today they needed the extra vessel because the musciara was bearing the saints.

The cross, its arms spread in submission, lay supine in the little gray wooden boat. All along its beams were extra-large holy cards. Two pictures of the Virgin and Child, in saturated reds and blues, bracketed the traverse beam. In between was a row of protectors: Saint Theresa, the Little Flower, holding an armful of roses; the Holy Family; the Holy Face of Christ; Saint Lucy, virgin martyr, proferring her eyes on a plate; and Saint Anthony in his belted brown robe. At the top of the vertical beam was Saint Ann, the mother of Mary; below her, Saint Joseph, Mary's husband, bearing a lily; then another of Saint Joseph holding the child Jesus; Saint Peter, who died a martyr on an inverted crucifix; and the Sacred Heart of Jesus. Below the saints was the seven-inch bronze statue of Saint Peter enthroned as pope. Above the saints were the palm fronds, the calla lilies and gladioli, and they were wilting.

The rais had waited for a day when the sea was calm to set up the saints. Today they would take their place above the bocca di nassa, the entrance to the tuna trap. The rais allowed me to watch the simple ceremony and assigned me to the *Marte*, his own vessel for the day. His men took up oars and rowed us out into the green bay. Angelo threw us a rope from the towboat, and Rosario tied it to the bow. The line tightened, and we were pulled with several other boats toward the trap.

Setting up I Santi is like putting the last stitch in a quilt, or the last brush stroke on a painting. The sea was smooth. Tension melted with the soft, breathing waves. At the bocca di nassa the boats formed a cross over the four cables that met at the trap entrance. Bows to center, they arranged themselves like the petals of a dogwood flower.

The base of the wooden cross is whittled into a sharp stake that the men must insert into a steel ring at the cable juncture. But on this particular day, the *Mars*, *Jupiter*, *Mercury*, and the rais's nameless musciara would not align around this node. There was the usual discussion and loud argument anytime anything is done by a committee of Sicilians. The rais smiled and joked. "What is this, the first time for you boys?"

With their boats finally in place, the men in the musciara set the cross in its base and lifted the saints. The cross rose up like the flag at Iwo Jima. It faced east, from where the fish would come. When it was straight and stable, the men removed their caps and stood in their boats for the prayer. It was the first time of the year the litany was recited, the last time that season the rais would say it himself.

"A Hail Mary to the Madonna of Tràpani!" From then on Rosario, the captain of the musciara, would shout the litany three times a day into the wind—twice in the morning going to and from the trap, and once in the afternoon on the return from the day's final check.

Their prayer was heartfelt and ended with the plea to Saint Peter to pray to the Lord for a good catch. "That he make it so!" they shouted in one loud voice. The comforting outline of Favignana's mountain hovered above us. We felt the downward gaze of Santa Caterina. This ritual was as old as the castle. The Arab Muslims who brought this method of fishing to the island in the ninth century did not pray to the saints, but Ruggero II, a Norman Catholic, came just two hundred years after them. He built his fortress a thousand years ago on the site of the old Arab watchtower.

The sun shone on the brief ceremony, and when it was done the rais went back to land on the towboat and we in the *Mars* tied up above the gate between the Camera Grande and the bastardella. The men untied the net gate and let it fall to the seafloor, and we took turns looking through the glass to count tuna, if tuna there were.

I was to look first and took my post as *specchiaiolo*. I spread out a burlap cloth, tied my hair back, and lay on my belly. The boat rocked like a cradle. Someone drew a tarp over the bench that I lay under, and the bottom of the *Marte* became my private place, me alone with my face in the seascape, bright and blue and filling me. I stared expectantly into another world. The sun's rays danced in the net. Thick bars of sunlight crossed and uncrossed in collimated beams, beautiful as the aurora borealis, and as moving. The net was a wall of fiber diamonds, stretching down beyond seeing, bulging with the eastern current. My hours on the tuna boats were among life's best.

"If you see a fish, say so," Clemente said. On my word the men would raise the gate and block the fish's retreat. Soon he joined me under the tarp, and we touched lips over the sea window. It was the only time we had been alone since Margo arrived. We watched the deep together, he on his side of the window, me on mine, our faces lit up blue. Then he withdrew.

Sometimes when they lower the gate to let new arrivals into the bastardella a few of the fish that were already in there escape to the great room again. "It makes you crazy," said Michele, the captain of the *Marte*. One of the crew members opened a little cabinet door built into the bow and passed around the white plastic sacks.

Bread, cheese, olives, and fruit. Somebody handed me a pear under the tarp, then took it back to peel it, then handed it back to me, cold and slippery. Only my feet stuck out from under the tarp. While I was down there one of the men asked me how old I was.

"Forty-one," I said. Another man said, "You shouldn't ask a woman how old she is," but not until after he had heard my answer from under the tarp.

"Are you married?"

"No. I'm too independent." One of my stock answers.

The men were anxious to look for fish themselves, so too soon my watch was over. They rolled back the tarp, and I blinked in the sun and wiggled out from under the benches. We talked, rocking quietly.

Michele got married when he was sixteen, became a father at eighteen. "If I had to do it all over again, I'd wait before getting married," he said. His hair was still black and curly in his middle age, and his wife's good cooking was beginning to show on him. "I've got a boy. He's fourteen, he's smart, but he doesn't like school." Compulsory education in Italy ends when a child is fourteen years old. That's when many young men quit school, stay home, get their parents to buy them a motor scooter, and proceed to burn gas and money. "I've told him if he doesn't continue school I'll make him get a job," Michele said.

Peppe, who must have been pushing sixty-five, was creased, wiry, and gaunt, with a large, permanent bump the size of a chestnut on his forehead. A happy soul, he talked incessantly, even while Michele was giving orders. Michele tried gently to persuade Peppe to shut up so he could enjoy the quiet sea. "If Peppe didn't talk so much, he wouldn't be Peppe," Michele said to me.

Paolo sat in the bow, smoking meditatively. He must have been in his fifties, a huge man with no front teeth who kept his own counsel. He had a nose that filled half his face. He said he understood my urge to travel and be free; he had spent many years as a crewman on cargo ships carrying grain and coal around the Great Lakes, to California, Florida, Virginia, Africa, China, Toronto, and Japan. "I was never home," he said. Once his ship was stuck in the Great Lakes for three winter months, trapped in lake ice. "I'll never forget that," Paolo said.

His buddy, Momino, was about the same age and just as small as Paolo was big. He lived the same life when he was younger. "Great memories," he said. "But I was away a lot. I didn't see my son until he was eighteen months old." His son, Anthony, sat beside him in the *Marte*.

This was Anthony's fifth year as a tonnaroto. He was married with two children.

"You could be a tonnarota if you wanted," he said.

"I know nothing of the trade," I said.

"I didn't know anything when I started," he said. "It takes just a year to learn what to do." Then the talk reverted to food and their stomachs.

"Tomorrow we'll take a little stove with us and make caffè latte and dip bread in it," Momino said. "In the afternoon we'll bring *amaro* to wash down lunch. You come too."

The towboat arrived, signaling the end of the watch. We had seen no tuna, but the men were still in a good mood. Michele stood with his hand on the rudder as we were towed back to port. The saints' cross stood outlined against the sun. Michele threw a backward glance at it, and at the two palms marking the trap's eastern end like lightning rods, conduits of energy, calling the bluefin.

"Ora, sì, che c'è tonnara." Now, yes, we have a tonnara. And with those words the waiting began.

16

Waiting

"Did you wash your face with flowers this morning?" Vito asked when I met him an hour late the morning of the First of May.

On that day the people of Favignana wash their faces with flowers for good luck. The night before everyone but fishermen picks poppies and buttercups and sprinkles the petals in a bowl of water. In the morning they cup their hands and wet their faces with the flower water, as I had done, first thing. But fishermen go to the water's edge and wash their faces with the sea.

Vito had invited me to an all-day picnic with his wife, two sons, and his brother. Il Primo Maggio was a national holiday, but Vito, who rowed for the rais, had already taken him to check the trap at seven that morning. By nine the sun was baking the damp chill out of the black marble slabs that paved the piazza. Chairs emerged on sidewalks in front of cafés. The wash on the line took less time to dry. There was a metamorphosis going on. The butterfly island emerged from its winter cocoon and unfurled its wings, the veins filled with blood. The sea itself was filling with the heat and light of life.

"We had one fish of two hundred kilos today," Vito said. Another tangled tuna. "The rais gave everybody in the boat sardines"—to roast on their holiday picnics.

The six of us squeezed into Vito's tiny car, women and children in the backseat, with bowls and small boys balanced on our knees. We headed to Il Bosco (the Forest), on the other side of Monte Santa Caterina, where we picnicked in a grove of pines, sparse but shady, near the bay where they say Odysseus washed up after his shipwreck. Vito's brother, Nicola, made a fire with dry broken branches while Vito's wife unpacked the artichokes she had stuffed with mint, olive oil, parsley, and garlic. She nestled them in the coals to roast while we ate rice salad dotted with firm, sweet peas. We dipped fava beans in vinegar. The men grilled sardines and beef cutlets. Around noon Nicola and Vito disappeared with the car and came back with two-liter soda bottles filled with Favignana's strong rosé, pressed from grapes grown in hidden gardens sunk in abandoned tufa mines, where the salty wind does not blow. We ate and drank, then took naps stretched out on blankets in sun-dappled shade. I woke up chilled under the long shadows and took the two boys for a walk along the shore. The wind blew up whitecaps. We packed up and left, and Vito drove us around the perimeter of the island. Near the lighthouse at Punta Sottile, flat and rocky and desolately beautiful, Vito spotted some soaring *spezzafelli*, seabirds that follow the tuna.

"Tomorrow we'll find some fish in the net." He didn't mean tangled up and dead, but swimming in slow circles, trapped, alive, and in love.

The trap was complete. Now it needed only maintenance work—patrolling the barrier nets for holes made by swordfish and dolphins, tightening the anchors, adding or replacing floats, the ritual counting of the fish. The night watch was set up to guard the trap against wayward craft. The attention of the whole town turned to the sea.

In the spring the trap becomes Favignana's soul, a hallowed, submerged court filled with a million cubic meters of dark blue water. The people feel its charge.

"How many fish?" is the question on everyone's lips. The islanders feel enriched by the presence of tuna in the trap, even those with no stake in the catch. The widows, the café keepers, the mechanics all ask, "How many?" The tonnaroti are not even allowed to make a guess. These days the number is a closely guarded secret. Raises have always been jealous of the trap, but in the past, when asked, a rais would give a number. The person who asked would know the rais could be fudging by fifty fish, more or less, but he got an answer. That was what one frustrated resident told me.

A *buon' annata*, a good year for the tonnara, was an index of fortune for Favignana, a measure of the temperament of the gods. But these days, with a single perfect giant bluefin selling for $60,000 and more at the Tokyo market, too much money was at stake to disclose the number. If the Japanese buyers got wind that many fish were in the trap, the price of bluefin on the world market could plummet. Only Girolamo, the diver, and the rais himself knew how many. The tonnaroti said they could lose their jobs if they mentioned a number. The people would have to wait until the day of the mattanza to know; not knowing left them hungry and unfulfilled.

As the days passed, then the weeks, the waiting got harder. The town was on edge. You could feel its pulse in the Piazza Madrice.

Years before, the Spanish fortress west of the piazza had become a regional prison whose inmates were now writing letters to the editor of the *Giornale di Sicilia* complaining about their lives in unheated caves. People clustered at their bars and clubs and muttered about the town's problems. There was talk of closing the prison; its guards, fifty local men, would lose their jobs. And Vito told me about a poster that had been plastered all over town in late winter, an open letter from the Parodi family to the people of Favignana, berating Castiglione for not having reactivated the Formica tonnara, defunct since 1980. The Parodis also blamed Castiglione for letting their own equipment rot—the black wooden boats, the old nets and rope cables, and the iron floats—while he brought in his new gear.

It was time for Castiglione to renew his lease, Vito told me, and he thought the poster was intended to get Castiglione to agree to better terms. The posters had all been removed after their allotted time in exposition, so I went to the town's poster headquarters to see if there were any left over.

A town worker handed me one, opened his refrigerator, and offered me a cold beer. The open letter was written in a baronial tone, black letters on Day-Glo red paper. It elicited the emotional support of the community. Because of Castiglione's inaction, it said, the people of Favignana had missed out on fifty new jobs that were not created, and eight months' work. The poster did not say what the work would have been, nor did it explain how Formica could be reactivated.

Once Formica had been a tonnara in its own right, manned by Favignana tonnaroti who spent the season in a dormitory there, living like monks (except for those who rowed for hours in the moonlight to sneak in a visit to their wives or lovers, Giaocchino told me). Formica once had its own rais, ciurma, trap, saints, and storehouses. The Parodis had sold the tonnara buildings on the uninhabited little island—so small it was named Ant—for use by a priest. He had turned its tonnara buildings, built by the Florios, into a treatment center for drug addicts.

The poster workshop was dark and cool, in the shadow of the high prison wall across the street. Once it was used to store tonnara equipment. Above my head soared beautiful pointed arches like those of the net house. Two dark framed oil paintings of women baking bread covered two walls. Mussolini, who exiled political prisoners to Favignana, had made this a prisoner workshop. Painted in Gothic letters up one side of the main arch and down the other was this slogan: "It is never too late to go further."

The church bell was tolling. A man rushed in, breathing hard. "Uno è morto," he said. Someone died. When he saw me, a stranger, he said no more about it. On Favignana the bell tolls as soon as the priest knows an islander has died, no matter where he has died. Women in black who never leave their homes have leaned out their windows to ask me, passing on my bike, for whom the bell tolls. Leonardo, who brought the bad news, drove the town's garbage truck. Seeing my open notebook, he sat down and lamented the fate of his island.

"L'isola non è più la nostra." The island is no longer ours. He longed for the days when nine thousand people lived here, when Favignana was self-sufficient. He remembered when Favignana exported honey, tufa, prickly pears, and cotton. Now half the population has emigrated to find work. He missed the days when men sat in tavernas and played cards and told stories and sang and ate octopus. He made me miss them too.

The cannery is closed, there is talk of closing the prison, and the tonnara employs half the men it used to. The tourists are taking over, he said. The island is being gentrified to death. The Favignanesi can't afford to live here anymore because tourists who buy vacation homes have jacked up real estate prices. "It will be an uninhabited island, with summer houses here for the rich from Palermo," he said.

During this period the tonnaroti got one day off a week. They were more relaxed, more likely to wander away for a coffee or an ice cream cone when the rais wasn't around. Often he was in Tràpani conferring with Castiglione or dealing with paperwork. The Camparia seemed peaceful, quiet. The tonnarotis' spirit rose and fell with the wind and the currents that brought the tuna. The *tramontana*, the wind from the north, was good, and so were the storms. But the weather was fine, too fine. Occhiuzzi sat in his undershirt idly unknotting a few cords of rope under the bell that is rung on the eve of the first mattanza.

"How many fish?" I asked.

"When the rais returned from the morning check," he said, "he reported two dead tuna."

"But how many are circling?"

Occhiuzzi kept his eyes on the rope. "I don't know," he said.

Gioacchino

The next day was Gioacchino Cataldo's day off. Still reeling from the rais's tongue-lashing at sea, he spent it painting the *Maria delle Grazie*, one of his two private boats. It was dry-docked at the main port near the ferry landing, a hundred yards from the Camparia. He took off his shirt and stood in the sun on the cement slip with two cans of paint, sea blue and orange. The paint spread like warm butter over the pale wood. The rough sanded boards gleamed under his brush. We talked for three hours.

Gioacchino pointed to his head. "I am never alone," he said. "I live with my mind. I've always got company." A conversation with Gioacchino was like a ride in a hot-air balloon: You started on the ground but quickly rose to a height where the world and the chatter in it became small and distant.

He connected to the world through his body.

"When I feel bad, when I kill tuna, I feel my heart move slightly to the right." He put his index finger on his chest and traced his heart's displacement. "Like a pain or a happiness. I feel it."

Gioacchino started work in the cannery when he was sixteen years old, but he didn't become a tonnaroto until the First of May in 1975, when he was thirty-four years old. He has served under three raises, Giacomo Rallo, Gioacchino Ernandes, and now Salvatore Spataro. When he hired on, all the nets were of coconut fiber.

"The tonnaroti of forty years ago, up until the 1960s, were poor people," he said. "As for their clothes, their pants were more patches than pants." He measured his financial success in terms of bread. In 1960, when he worked in a Milan factory, he earned 1,150 lire for 8 hours' work, enough for 10 loaves of bread when it cost 120 lire per kilo. Now, working 8 hours a day for 53,000 lire, he can buy 25 loaves of bread with a day's wages. He wrote out his bread formula in purple ink in my notebook. My pen looked like a toothpick in his hand.

"The world has changed," he said. I shifted position so I wouldn't have to squint into the sun. The bright orange paint in the inner curve of the bow reflected a fiery orange glow onto Gioacchino's torso and face, as if he were bent over a flame. When he was growing up in La Praya, the fishermen's neighborhood near the anchor beach, "there were people of other professions, the stonecutter and the farmer. Each took a turn at being the richest on the island," he said.

First the stonecutters made a decent living, until transportation costs made Favignana's tufa too expensive to export, then the farmers who made honey and raised hay, fruit, and cotton. Now it is the fishermen's turn, with fish prices high because of the scarcity of the product.

I asked him how many people lived on Favignana. He had me write down "3,539," then subtract 80, the number of people who had left since the last census. There were 3,459 people on the island now, he answered precisely.

He then asked me, "What is the most beautiful thing, the strongest thing, and the most cutting?" I had no idea.

"Love, death, and truth," he said.

He commented on how well basil seed grows on this island, but noted that in Germany the basil has no fragrance. Then he wrote a poem in my notebook:

Alcune volte un sincero
Grazie detto con il cuore
E con la luce degli occhi
Può rendere felice un'uomo.
(Sometimes a sincere
Thank-you said with the heart
And with the light of the eyes
Can make a man happy.)

Then he signed it. Gioacchino Cataldo. He had composed it years before, one day at sea. He keeps a book of forty to forty-five typewritten sheets of beautiful sayings on his private boat, he said, quotations from the most important authors and thinkers of the world. He inserts this poem in the middle of the book and asks the summer tourists he takes for boat rides to choose their favorite saying. The people choose his saying "83 to 85 percent of the time"—"people who have gone to college, professionals." Gioacchino made it only through fifth grade before he went to work.

As we spoke, the coffin bearing the body of Tony, a young man who ran a pizzeria with his German wife and their children, was being unloaded from the ferry. He had died in a Palermo hospital the day before, when the bell had tolled, from head injuries sustained on Favignana. They say he was on his motorcycle on the road to Punta Lunga, fell, and hit his head on a tufa wall. Now he was coming home to be buried on his island. The funeral van, covered with gladioli, arrived to meet the coffin.

We watched as all the trucks, cars, and passengers disembarked first. A high sea rocked the ferry and made it difficult to unload. Townspeople gathered to meet Tony's body. The white ship blew three long farewell blasts, then came three blasts from the hydrofoil *Donatello*, then three from the *Vulcano*. The stentorian blare vibrated in my ribs.

The horns wailed because Tony was so young, because he loved the sea, because he had a wife and children, because the Fates had cut his thread so short.

"What should a man give a woman so that she will not forget him?"

Death did not deter Gioacchino.

"I'd give her a skein of coconut cord from the old tonnara," I said, hoping.

"No, because you could lose that."

I gave up. "Jewelry?"

"No, because you could lose that too. I'd give her a kiss that she wouldn't forget."

The plain wooden coffin was carried off the ferry. Six men carried it on their shoulders with the van following behind. In front was Vito, holding on to the shoulder of Carlo, who had knifed him the year before. The church bell tolled. The priest waited at the top of the pier with two altar boys. Some four hundred people followed the coffin in silence. The women made the sign of the cross over and over. The bicycle shops at the port lowered their shutters as the coffin approached and raised them again after it passed. The pallbearers disappeared around the corner on their way to the piazza.

Gioacchino went off into the wild blue yonder of his mind. "I hated Favignana when I was growing up. It was like a prison. There were two cinemas here then. I could see things that were very taboo. There was no love, no sexual adventure here. So I went away and found everything."

He went to Germany and found Teutonic goddesses, the Valkyrie, he called them.

"What is your worst memory of Favignana?" I asked.

He told me the story of one Pasquetta, the day after Easter, a national holiday in Italy. On that day Italian children get Easter eggs baked into hard, sweet pastry crust. One year Gioacchino, with a wife and two children, had only five hundred lire to his name, not even enough for a pack of cigarettes. (He used to smoke two packs a day, but he quit.) The worst was when his children asked him, "Papà, will you buy me an Easter egg?"

"As soon as I can go fishing I will have some money," he told them. So that Pasquetta, instead of picnicking with his family, Gioacchino went fishing, caught eighteen kilos of little squid—*calamaretti*—worth about a hundred dollars today. He sold them, collected his children, and took them to a caffé where there were still plenty of Easter eggs for sale.

"Choose any eggs you want," he told them. "And they chose small to middle-sized eggs," he remembered.

Gioacchino is the greatest repository of mattanza statistics there is. Even the rais, when I asked him for the tuna catch numbers of years past, advised me to ask Gioacchino. As he brushed blue, his favorite color, "the color of the sea and the sky," onto the curve of his bow, I asked him for the counts.

"In 1864 there were 14,020," he said. "That's counting only Favignana, under Rais Casubuolo. In 1957, 7,500 tuna." He straightened up. "Turn the page," he said, commandeering my notebook. "Make a list down the page, 1974 to 1993"—a line for each year. These were his years as a tonnaroto. He went through the list slowly as he painted. His look was not of concentration but of a man transfixed. I could see him travel back to every year, remembering the color of the sky, the shirt he wore, how long his sideburns were in each year. I called out the year, he came up with the number:

1974 . . . 2,500
1975 . . . 3,083

1976 . . . 2,500
1977 . . . 1,500
1978 . . . 900

"Draw a line there," he said. "That was the last year Favignana combined with Formica."

1979 . . . 640
1980 . . . 502
1981 . . . 650
1982 . . . 750
1983 . . . 850
1984 . . . 440
1985 . . . 660
1986 . . . 1,036
1987 . . . 285
1988 . . . 390
1989 . . . 1,354
1990 . . . 1,336
1991 . . . 1,279
1992 . . . 1,489
1993 . . . 977

I was amazed. "Did you memorize the list?"

"No. I carry it in my heart."

He didn't need to memorize the numbers any more than I needed to memorize what I was doing when Kennedy was shot. Gioacchino loved the tonnara, every cable and knot, oar and anchor, the currents and winds—it was all engraved on his heart.

"Have you heard the rumor that Castiglione plans to eliminate the mattanza and instead just wait for the tuna to die in the water?" I asked him. A crane would pluck them from the Chamber of Death so their flesh would not be bruised by the bite of the gaff.

"I hear that rumor every year," Gioacchino said, "but no one has seen the crane." To him the mattanza was integral to the tonnara.

"Even though it hurts my heart," Gioacchino said, "the mattanza, the slaughter, is for me like a plant I planted. After a period of time I take the fruit. But it must be me to take the fruit, not a crane. I release my nervous charge of the entire year on the poor tuna. When I finish a mattanza it's like I took flowers to the cemetery. I feel calmer, more relaxed."

Then he went home to change into his church clothes before attending Tony's funeral.

18
Tufa

The thick tufa walls of my house were coated with lumpy oatmeal stucco and whitewashed inside and out. The outer walls shone. Inside was the cool feel of a cave. My bedroom was a stone box of light with no adornments. The first thing I saw each morning was the pink-orange rectangle the rising sun projected on the wall; at dawn it hung there awhile, a nubby tapestry of light, then slipped to the floor and disappeared before seven. I would shuffle into the kitchen and brew a pot of espresso, pour it into a bowl of hot sugared milk, and lean out my back window to look at the day.

A deep, played-out tufa mine spread from the foundation like an empty moat. The quarry walls rose sheer and square, washed in peach and pink, crosshatched with cutters' marks. Wild capers grew from the cracks, their blossoms white trumpets blowing sprays of violet stamen. Birds nested in niches that the caper roots had crumbled. Seagulls swept overhead; white calla lilies grew under a juniper tree in the garden.

Michele Ingrassia, the owner, took care of it all. A methodical, punctual man, he lived in town and came every day but Sunday to water the flowers and work on his retirement project: cutting a tunnel through the column of tufa that held up this house so he could link the front and back quarries. "Stonecutting is in my blood," he said. Everybody's a sculptor on Favignana; they can't keep their hands off the tufa.

The first person Piero and I had met on Favignana in 1986 was the naif sculptor Rosario Santamaria, known as Zu Sarino. He must have been in his seventies, a wiry guy with a sharp, pointed face and lots of energy, who rode around the ferry quay on a tiny bicycle, his knees practically hitting his chin. He wore a sign around his neck that asked arriving tourists: "Please keep our island clean." In the morning we'd see him down at the Camparia boat dock, raking in flotsam with a boat hook. A sticker on his bike said he was a member of the World Wildlife Fund.

He was a self-trained artist, and tufa was his medium. He had spent his life in the quarries when they were vibrant and men chipped the blocks out by hand and loaded them onto flat boats with sails. He looked chiseled himself. He'd retired from quarrying, but he still liked chipping rock, so he biked around the island and carved Modigliani faces in courtyard walls. He always cut his initials under his work: "R.S." No one complained; Zu Sarino's carvings were a mark of privilege.

The ten-foot-tall sitting queen in San Giuseppe Bay was Zu Sarino's work. He had shaped her so that the sea washed her feet. His house in the old quarter was a riot of color that stood out in a street of plain white facades. He had a habit of collecting old painted majolica tiles from demolition sites and plastering them to the exterior of his house, which was now covered in deep blues, greens, oranges, and reds.

A local engineer who owned a building near the Camparia let him use it for his artwork, free of charge. Zu Sarinu, free spirit, set up his studio in the courtyard where passersby could watch. He would not sell his work but gave the tufa back, transformed, to islanders and friends. Zu Sarinu turned out stone moons, stone cats, and stone baskets that he nailed to the front of the house. Along the low wall he perched his three-dimensional work: saints' shrines, flower pots, stone ships a foot tall with chunky tufa sails, and giant primitive busts with the mad stare of the gods at Easter Island. I'd seen such busts topping pillars at the ends of driveways all over the island. He asked me which piece I wanted.

"I'd like a tufa tuna, but I don't see any," I said.

Zu Sarino palmed a flat chunk of rock, chiseled its sides smooth and flat, cut the profile of a bluefin with a bandsaw, a torpedo with a lunate tail and gaping mouth. With a hammer and screwdriver he dinted its flanks with the essence of mottling. Two gouges for eyes, a line for the gills, then he handed it to me, a synthesis of Favignana's two cultures, tuna and tufa.

The stone had apparently given Zu Sarino his health, a sense of humor, and an unbridled imagination. I have pictures of him jumping rope, raking the sea clean, his watch cap pushed over one eye, bony knees poking under his chinos. In one shot he's stretched out in a gutter, playing dead, next to a curbstone where he'd carved his full name.

Most mornings I'd jog ten minutes from my house to Cala Rossa, where I'd find petrified scallop shells embedded in the rock and round pincushion skeletons of petrified sea urchins bigger than my hand. I'd stand at the cliff edge and stare down into the gem-blue water and the sea foam that limned the cliff base, then jog home feeling like a vibrating crystal, even on days when I met the old shepherd who hailed me with obscene masturbatory gestures and asked me in a hoarse whisper, "Che fai? Che fai?"

Fifty yards past my gate the dirt road led to Bue Marino, the Sea Ox, a flat shelf of tufa rock under a quarried cliff where sea lions once sunned and now so did I. About every ten days, when I needed a break from the tonnara, I'd come here with a pillow under my arm, stretch out on a rock shaped like a fainting couch, and read Homer. Tiny caper blossoms sprouted from the fissures in the sofa. Two-inch-tall purple iris grew in the sand and popped up, exquisite, from bare yellow rock.

Some scholars say Favignana is where Circe turned Odysseus's men into swine. Others say it is the island of

Ogygia where the nymph Calypso rescued Odysseus from his shipwreck, "drifting alone astride the keel of his ship." She became enamored and kept him in her love cave for seven years. When Hermes showed up bearing Zeus's order to release the hero, the messenger found her cave ". . . sheltered by a verdant copse of alders, aspens, and fragrant cypresses . . . ; and in soft meadows on either side the iris and the parsley flourished. It was indeed a spot where an immortal visitor must pause to gaze in wonder and delight."

Before me spread out the blue sea. Across the strait was the northwest coast of Sicily, with Erice humped over Tràpani. Behind me were the forty-foot walls of a tufa cave, one of the oldest quarries on the island. Bue Marino was where cargo boats once loaded the stone blocks for export; the old stone chutes were still intact.

Sometimes I'd stay all day and read. In midmorning a fisherman would tie up at Bue Marino, turn on his radio, and sort his net into piles, one with holes, one with fish. Sometimes at noon I'd look up and see a tonnaroto peering over the rise, come to check on me because I hadn't shown up for work.

At night from my patio I could watch the flickering lights of Tràpani and Marsala curve up the Sicilian coast. Once I climbed up to the roof terrace to watch fireworks burst in silence over Marsala and drop like spiders; the booms came minutes later. A fishing boat cut a white wake in the glassy black sea. The foam glittered with phosphorescent green sparks. Back in my room the moon cast a blue rectangle on the wall, and I fell asleep under the steady sweep of a lighthouse beam.

When Margo arrived with Clemente's daughter I saw much less of him and much more of his son Angelo at my table. I wondered if Angelo liked my cooking or if his father had sent him to amuse me by proxy. I tried to think like the Italians. Margo was Clemente's signora, the mother of his youngest child, and I was his mistress, both time-honored roles in Sicily. But it was useless; here was a culture gap I couldn't leap across.

One day Angelo had the afternoon off and invited me to watch him scuba dive at Punta Lunga, where old-timers keep their boats and nets. It is a small cove protected by a tufa seawall and furnished with two piers. The pensioners told me that before the new port was built on the north side of the island the Punta Lunga fishermen used to meet the ferries here and bring passengers and goods to shore. Now when the sea is rough the ferries don't come at all and the island, once self-sufficient, is left in the lurch.

One recent winter the island went eleven straight days without seeing a ferry, so the butchers ran out of meat, the bakeries ran out of flour, the newsstand racks were empty, and no one could come or leave. Even the town's drinking water is imported from Messina or Naples, and in bad weather the town can run dry.

Punta Lunga is home to a score of scrawny cats, mostly orange tigers, who rummage in trash bins for fish bones. A one-eared cat with runny eyes and a blue collar licked its paws complacently. "That's Ruggine." Angelo introduced me to the rust-colored cat while he pulled on his diving gear. "He's fourteen years old. All the cats you see around here are his children. There are hundreds of them." He said it like he was proud. After a while I got tired of watching his bubbles and went home to eat.

People on Favignana worship cats. On my evening ride into town I sometimes passed a man in a fedora and dark suit ladling pasta from a pot in the back of his car. He was Il Professore, a retired school principal who cooked the pasta for the cats and served it on paper plates from the back of his Ford hatchback. Precisely at 6:00 P.M. every night the cats perched on Dumpsters and waited for him. The Favignanesi get possessive about the feral cats they feed, and I'm told some resort to cat stealing. The cats are diseased, fully sexed and unvaccinated, but on Favignana they get fed.

I liked living fifty yards from the sea and five kilometers from the piazza, and I settled into a nice routine. I knew every bump, rock, hole, and incline on my route to town. I braked for the bright green lizards that skittered across my path. In the early morning two solitary caper pickers filled their white plastic bags with the pebble-hard buds they would pickle in brine. At the paved road I passed three cows plodding back to their barn, their herder on a bike a quarter-mile behind them, dodging fresh cow plops steaming in the chill air.

I always met an elderly fisherman with a basket of fish on his handlebars. Every morning I would say to him, "Buon giorno." At first he seemed not to hear; then for a couple of days he looked at my face as he passed but said nothing. One day he rolled to a stop. "Excuse me, Signorina, whose daughter are you?" In Sicily no mature woman would be caught dead on a bicycle shortly after dawn every day.

"I'm not the daughter of anyone here. I'm a tourist," I said. After that he always tipped his hat.

Night air on Favignana is damp and cool; I drank it like spring water every evening when I left the piazza for the half-hour ride home through the dark. I turned my back on the mountain, the din of motorbikes, and the talk in bars and crossed a stone bridge onto a path that was half asphalt, half bedrock. It led to a lane bordered by tufa walls, some jumbled and overtaken by capers, some smooth and fitted, the work of a seicento stonemason. The moon drained the colors from the landscape but not the fragrances; jasmine and pine swirled on the breeze, and the dry perfume of cut hay rose from fields where a month before red-orange poppies had grown.

The only sound was my front tire kissing the ground. The world was hushed when I reached my gate. I could hear the sea sigh behind my house, veiled in blue light. There was no one, no one. I was alone at the far east end of Favignana. No neighbors, no radio, no television, no phone.

One night I fell asleep in a moonbeam and dreamed of colors, an open box of crayons. In the morning the sun shone on me first.

I made a caffè latte and wrote up a list of things I ought to do. I still hadn't seen the inside of the Florios' Belle Epoque palace, which Favignana maintains as a memorial to its departed benefactors. Cola Pesce, its custodian, was from Marettimo, one of Favignana's two sister islands. He took his nickname from the legendary boy who loved the Sicilian sea so much his mother called him Nick Fish. He could dive deeper and longer than anyone. One day the king asked him to find out what kept Sicily afloat. Cola Pesce dove and found three pillars sustaining the island. One was broken; he took its place, and he's down there still. I stood at the palace gate and called his name.

Cola Pésce showed me around. The rooms were grand and empty, but the Florio numen still hung in the air near the high ceilings. A polished banister curled up a marble staircase lit by a stained-glass window emblazoned with the Florio coat of arms, with the crouching lion on a three-cornered island. The upper rooms held an exhibit of the artwork of local fifth-graders whose teacher had given them a poem by Fadwah Tuqan, a Palestinian poetess, to illustrate. One drew a box of crayons. One drew Monte Santa Caterina. One drew fish in the trap.

Island where our dreams dream
Let us leave at last.
Free us from your seduction,
Absurd mirage of light
Web of unseen threads
Which caught us, like a net,
To hurl us to a desert.
Island where our dreams dream,
Cause of our perdition.

It could have been the voice of Favignana; it could have been the song of the bluefin; it could have been my song. The island had cast a spell on me; the tonnara was the island's own dream, a gorgeous chimera anchored in a sensuous landscape. I cried on Cola Pesce's shoulder when I thought of having to leave this place. I didn't want to wake from the dream.

"You'll be back," Cola Pesce assured me, and led me to a porcelain basin where I could splash my face with cold water. Then I sat on the palace steps facing the sea and memorized the poem I'd copied in my notebook.

After that day the islanders were extra tender with me. Benito Ventrone, the tonnaroto, sewed up a tear in my jeans. At the Two Columns Stefano kept filling my wineglass, on the house. Ninetta, who worked at the bar, brought me a rose, its white petals rimmed in red, and told me to press it in a book.

When the islanders saw me, one condemned to leave, their own deep feelings for Favignana began to surface.

Antonio Casablanca took me to a secret spot in a dense, tangled mulberry copse where he had raised rabbits in a tufa pit as a boy. "In the summer my friends and I would eat as many berries as we could, then strip to our shorts and smear the juice all over us," Antonio said. Thus anointed, they ran howling into the piazza to scare their mothers into thinking they were all bloodied, or daft.

The ciurma loved me, Antonio said, everybody wanted me on their boat.

Every morning in the spring, before she left to open the Two Columns, Cristina parted her kitchen curtains to watch the tonnara boats go to sea. She told me that once, when she was about to have surgery, just before the anesthetist placed the mask on her face, when she didn't know if she'd wake up again, she closed her eyes and saw the tuna boats being towed in a line.

Antonio Noto, my photographer friend, got his motorcycle out of storage and took me to a shallow inlet near Il Faraglione, the western lighthouse, where his cousin used to bring him to bathe when he was a toddler. It was a pocket of white sand under clear green water between two fingers of bedrock, a memory from before he had words. As an adult he made the lighthouse the subject of his most beautiful photographs. We set up our tripods and waited for the sun to sink. We shot it all through the long twilight, every few minutes, a black silhouette under the red sky, under the orange sky, under the purple and indigo.

ꝏ

I spent most days following the tonnaroti to work, to bars, to their boats. Small, lazy pleasures became precious.

Vito Giangrasso didn't know what to do with himself on his day off, so when he showed up at the Camparia just as I arrived at seven, he invited me for a tour of the western side of the island. In the car I asked him how he felt during the mattanza when the rais gives the order to kill.

"In that moment it's like you're a hero, like the hero of a movie," he said. "The people are watching. It's a spectacle."

Every rock, dip, rise, curve, quarry, cave, and clump of trees on the island has a name. We drove to Il Pozzo (the Well), a jagged outcropping where the surf pounded to shore. There we found Benito, his white hair in a ponytail, squatting on his haunches against the wind with a fishing rod in his hand. We squatted beside him and watched his line awhile. The surf broke against a shelf of tufa; the sunlight slanted through the spray and made a rainbow.

I was hungry by midmorning when Vito dropped me off in the piazza, so I went to La Madonnina bakery next to the little Madonna in Via Roma for a *pizzetta*, a palm-sized pizza. I found Domenico, the redheaded baker, snapping an old guy on the chin.

"Why?" I wondered aloud. Domenico explained that old Signore Ponzio used to work the tufa mines when the rock was dug by hand, before machines did the cutting and lifting. Signore Ponzio wore thick glasses and had gray stubble on his hollow cheeks. He had proffered his pointy chin for abuse in a tough-guy demonstration of his high pain threshold.

"He can break a shot glass with his index finger," Domenico said. "He used to win bar bets this way."

"I don't bet anymore, but I'll show you what I can do," Signore Ponzio said.

Thwack! He rapped his finger against the seat of a wooden chair. It rattled then settled. *Thwack!* against the glass counter, making Domenico nervous. On his way out he rapped the door glass, holding back so he wouldn't break it. I followed him out.

"Is your finger numb?" I asked.

"No, I still have feeling in it," he said, "but the calluses dull the pain."

He held out his hand for me to touch, and it felt as tough as tufa.

19

Hot Air

When the sirocco blew up from Africa, it blew for three days. The humid air settled over the island like a leaden haze and pressed. It gave me a sinus headache and put everyone I met in a glum stupor. At the bank the teller who changed my money yawned; she'd been yawning once a minute, she said. "E lo scirocco." People in the piazza sat at tables with their chins in their hands. In the back room of the Two Columns I found Clemente dressed in purple rags. He was slumped in a chair and groaning.

"Is that what you were wearing when the nun ran you down?" I asked. I reckoned today was the two-week anniversary of his accident and he was commemorating, or just reminding everybody. The first mattanza had to be coming up soon. "How are you going to be able to kill tuna in such pain?"

"It's like making love," he said. "When you're in the middle of it, you don't feel any pain. You just need to get your blood heated up. You hurt later."

I couldn't work up any sympathy for his pain, and he wisely clammed up. Everyone looked like they had hangovers; it was depressing. The nut vendor said the piazza had been empty last night—the sirocco was even affecting business. Sicilians are sensitive to invisible currents. Back in Santa Margherita Belice, my ancestral village, when I was gasping for air one stifling summer night, my old cousin Calogero insisted I shut the window or the "malaria" would get me. As soon as he turned his back I opened it again and breathed the evil night air.

At the Camparia the next day Clemente took me into the anchor maintenance workshop. I saw him nowhere anymore but in the workplace. In the dim light he removed the lid from a small barrel of salt and pulled out a ten-inch-long swordfish egg sack. A delicacy, it is dried, sliced thinly, and eaten with a little olive oil. He put his fingers to his lips; no one else was to know about his stash. I thought of a line in "Gnazù," one of the ancient work songs, that mentions the *rubalattumi*, the egg-sack stealers.

The rais was gone to Tràpani. The tonnaroti were horsing around like teenagers in the net house as they attached five-pound terra cotta weights every eighteen inches along the bottom of the *ancera*'s seventy-foot width. The ancera is the coconut fiber net that waves in the current like a blowing curtain to scare the tuna from the bastardella into the Camera della Morte. The pink weights have a technical name, but the tonnaroti call them *coglioni,* a vulgar term for testicles.

"Dammi acqua!" the men yelled, calling for more twine.

Familiar voices echoed off the soaring arches. By now I could put a face to almost every voice. Each had a life and a story. Some were stonemasons when they weren't trapping tuna. One was a truck driver who had lost his license and become a tonnaroto to pay off his fine. Many had crewed on oil tankers and cargo ships and gone around the world. Many were sons of tonnaroti. Fafarello, who loved the anchors, used to work in a mine in Germany, hundreds of meters underground, to make enough money to get married. For a man from this sunny island a trip down the shaft must have been a descent into hell. "Una bella vita," he said. "Every time I went down I made the sign of the cross."

Swallows flew in and out of the rafters. A cone of light beamed through the clerestory, my eyes grew accustomed to the dark, men came in from the sunshine.

Almost everyone on the island had a strange nickname, inherited from their fathers. There was Cipolla Rossa, the Red Onion. Occhiuzzi was Small Eyes. Guardamadonna's home used to "look at the Madonna" near the statue of the Virgin in Via Roma. Clemente was Scaminali, the term for the smallfry inadvertently caught in the trap. The men had a good laugh when I told them mine—Passaluna, which in my grandfather's village meant "black olive" but here on the island meant "overripe fruit."

To everyone's surprise, Vincenzo Sercia, the former First Voice, walked into the net house one morning. He had been a tonnaroto from 1947 to 1991. He apparently had had a feud with the rais and never came near the place. His legs were short and bowed, his eyes rimmed with red. He wore his knit cap tilted jauntily over one eye. He knew all the verses to the *cialome* and used to sing them during the mattanza. I asked him how he learned all the words. Everybody crowded around to listen.

"When I was a kid, there was just a sloping beach in front of the Camparia," he said, instead of the flat macadam platform there is now. "The men used to load the nets there. I would stand there and listen to Peppe Trumba, an old guy, the Prima Voce before me. I learned 'Ai-a-mola' without ever writing it down."

Sercia was seventeen when he entered the tonnara. Gioacchino Ernandes was rais, and back then it was the rais who sang the songs. Then, when Sercia was twenty-one or twenty-two, the rais told him, "You can sing now."

"There were good years and bad years." He remembered the year of the great white shark, found in the trap on May 29, 1952. They had to bring a *carabiniere* to shoot it with a machine gun.

"We used to be seventy-three tonnaroti, and twenty-four *avendizzi*." These were auxiliary workers called in at the end of the season to help pull up the nets and bring in the anchors.

The men begged him to sing for my tape recorder, but he refused. He said he needed the rais's permission to sing in the Camparia.

Gioacchino had been telling me that he kept "something good for mankind" in the medicine cabinet at the Camparia. I guessed it was a bottle of wine, but he said no. Someday he'd show it to me. Now he went to get it. The men who had wandered away to the dockside bar suddenly wandered back in. Gioacchino brought out a wooden container the size of a shoe box painted red with a white cross on it, like a first aid kit. He set it down on a workbench and invited me to unlatch it. The men crowded around three deep.

"Don't worry, nothing will spring out at you," Gioacchino said.

When I unlatched it a giant wooden penis, painted red, sprang out at me. A cheesecake magazine photo was pasted to the roof of the box. Guffaws all around. If I laughed I was a wanton woman. If I disapproved I was a prude. I stood with a wan smile on my face. "I should have known," I said. I must have struck the right balance.

"What's the lightest thing in the world?" Sercia asked the company in general. "A penis! Because it rises on its own." Carlo elbowed me in the ribs. "He knows thousands of them." I recorded a half-hour of Sercia's ribald jokes, but he kept them "light" for my sake.

I spent a lazy hour by myself on the southern coast in a little cove where the sandstone cliff curved over me and the sea washed up into little pools on the flat bedrock. Then I biked back to the Camparia. Capoguardia Messina was there with some of the men, idly tying knots in a rope. When I walked through the gate he greeted me.

"Auguri," he said, offering me best wishes. It wasn't my birthday.

"Why 'auguri'?" I asked.

"Because you're a rose, and not a rose of summer, all withered, but a rose in winter." They had been drinking. They kept inviting me into a small storeroom—someone had penciled "TAVERNA" on its door—for a furtive paper cup of Favignana wine. But then we brought our cups out into the sunshine.

Messina sat in a coil of rope and handed me my third drink. When I sat down in a coil next to him Gioacchino got a pained look on his face.

"Don't drink it if you're not used to it," he said. They were looking out for me. I handed the cup back to Messina, who downed it in one gulp. Auguri.

One day, for a change, I biked over to Punta Lunga, where I found Rosario, the bow man for the musciara, tending to his private boat on a day off. He was named for Our Lady of the Rosary, the fisherman's madonna, whose feast is in October.

Cyclones have a tendency to form in the waters on this side of the island at that time of year, he said. In Sicily a cyclone is called a *tromba marina*, a "sea horn." "One touched land just a few years ago and wrecked an empty house near the port," Rosario said. He pointed to another house the cyclone had stripped down to its ancient stone arches. "Fortunately no one was hurt because it was a holiday and everyone was home eating or napping," he said. Otherwise, "this place would have been full of fishermen" readying their boats and nets.

Rosario knew of two ways to stop a cyclone at sea. "If there is anyone aboard born the first Friday of March, he is to lower his pants and moon the sea horn," Rosario said. "Ship captains will always try to recruit one crew member born on that day." The other method is more complicated. "A man must learn a certain incantation at midnight on Christmas Eve, when the Child was born," Rosario said. He had learned it. The secret words are to be recited three times to the tornado "while making the sign of the cross with your hat on upside down and backward. It will disappear."

20

White Magic

My Sicilian grandmother never exhibited any knowledge of witchcraft, but it was certainly practiced in her hometown. In Santa Margherita the secrets of magic zigzagged across the sexes from generation to generation; a man could teach the art only to a woman, and a woman only to a man. Furthermore, the secrets could be transmitted only on a Good Friday, or from a deathbed. This I learned in Santa Margherita from a woman who had been initiated by a *mago* one Good

Friday. When her children were ill he would write a charm on a scrap of parchment to be worn in a leather bag about their necks. His fee for this service, a full sack of flour, was so expensive she decided to learn white magic for herself.

Sicilian wizards are pervasive. Their cheerless, omniscient faces stare into the middle distance from posters plastered at bus stops all over the main island. They drive through Palermo's back streets and fling handbills by the fistful. They flutter around like Chinese menus in New York. I collect them. Madame Sahara gives advice to the lovelorn. "Consultations by mail. Send 30,000 lire." Or my favorite, the Mago Giusa, with his Geraldo Rivera mustache: "Prepares talismans on parchment and precious metals. Protects against enemies, curses, and sickness. The strength of his mind knows no bounds."

They have their own television shows. I watched one at Rosetta's house—a lady reads the tarot cards on the air. Welfare mothers dial her 900 number and wait a sufficiently profitable amount of time before being allowed to ask their free question.

It's not just the poor or uneducated who resort to magic in Sicily. Instead of consulting a marriage counselor, upper middle-class spouses who suspect their partner of infidelity pay a minimum of four hundred dollars to have a curse put on the interloper, or more benignly, to have the errant one brought back to the nest. One popular remedy for male infidelity is for a woman to collect her menstrual blood, dry it, and sprinkle the powder into her husband's pasta every night for a month.

In one town near Palermo the holy water fonts in the church were kept dry, a friend told me, because the priest had learned that practitioners of black magic had been stealing the water for their evil purposes. Communicants at mass were not allowed to receive the consecrated hosts into their own hands because they too had been pocketed and used in black magic rituals, she said.

Some of the women on Favignana practice white magic. One spring, when renting a house was no longer an option, I found myself on Signora Neri's doorstep, looking to rent a room. Her freshly whitewashed house was in the warren of narrow twisting streets behind the post office, close to the port. I pressed the doorbell, and Signora Neri opened the door. She was in her sixties, built like a snowman, a round head on a round torso on round hips, and short. She showed me in.

The entrance room was a sparkling dining room, with a massive, glass-fronted credenza full of china and wineglasses never used. A polished wood table with a crocheted lace cloth under thick, clear plastic took up most of the room. Behind it was the kitchen, where the real living took place, with a small table covered by an oilcloth and a bowl of brown bananas. The room she had for me had been a triangular walled courtyard until they roofed it. A foldout bed occupied most of the floor space; crammed under it were boxes and bags containing somebody else's life.

The room was infested with stuffed animals, encircled by shelves of dusty dolls, and closed in by three walls plastered with framed greeting cards and magazine travel ads, everything just as her daughter had left it when she married and moved out. I fought claustrophobia. The incubus came with clean sheets, meals, and use of the kitchen. I missed my house at Bue Marino, but at least now I would eat regularly, I'd be close to the Camparia and piazza, and I could afford it, so I took it.

Signora Neri opened the two dresser drawers she had emptied for me. "If the police ask you where you are lodged, please say you are my guest, a relative from America," she requested. Otherwise she'd have to pay business tax. "And please don't answer the telephone."

We were at her kitchen table. She was already knitting me booties, pink, with tasseled drawstrings. She muttered and fussed while she knitted. I found I could understand her quick Favignana dialect. She had a lot to complain about, and I spent hours listening to her.

She had married at twenty-seven, very late for a woman of her generation. Her husband was *lampadista* on a commercial fishing boat, the man who held the lights that the schools of fish swim to. The crew would be gone for weeks, sometimes to waters off Tunisia. The couple's engagement had lasted nine years while her sisters, both older and younger, were married off first. Signora Neri had to wait for her wedding "because my mother wanted to keep me as a servant, and she would not give me a *correda*." The traditional dowry consisted of the sheets, blankets, pillows, nightgowns, and furniture she would bring to her marriage.

Her betrothed took things into his own hands: He worked and sent her money to put aside for the correda. Finally they were married, and everything in their house was paid for. She said he never let her lack for a lira.

"I was the little lady of the house, and even did charity work," she said. She used to carry bags of groceries—oranges, olive oil, and bread—to her sisters and brother when she had a little extra. After twenty-eight years her husband died, and she had to move into her deceased parents' house. Now she was in need, and her older sister could give her only some fava beans from her garden or a couple of oranges when Signora Neri visited to give her injections. "You must never answer the phone because it might be my sister calling, and if she knew I was taking in boarders she would think I'm getting rich on our inheritance," she explained.

Her sister had an unmarried daughter who helped with the housework, but Signora Neri had no help, just an ache in her gut from worrying about her daughters, her bills, and her daughters' bills. "Doctors are no help," she said. She would not be seen walking alone to the doctor's office, and house visits cost seventy thousand lire, "to have your stomach thumped," so she stayed inside and suffered. The only time she went out was to seven o'clock mass every morning. The rest of the day she stayed home to clean and cook.

She put a pot of espresso on the burner and clicked on the radio. It was tuned to Radio Maria, the Vatican network. While she swept, scrubbed, and ironed she listened to religious music and call-in priest shows. She talked to her five cats, Cesaro, Geri, Kiku, Nicca, and Niccareddu, a scabrous, filthy, inbred crew that marauded through the house and slept on my bed if I didn't latch the door. Cesaro had learned how to open the refrigerator and now sat perusing its lighted contents.

Mornings in May, the month of Mary, Radio Maria broadcast a rosary recital. Signora Neri would put down her work to finger her beads in her lap. She never joined the women who recited the rosary and sang to the Madonna of the Tonnaroti every day at four. At noon, when she heard the bells of Saint Peter's ring the Angelus in Rome, it was time to set the table for lunch.

Signora Neri was a devotee of Santa Rita da Cascia, who blesses roses on her feast day and grants a wish for each rose. One morning Signora Neri gave me a rose and brought me to the old church of Sant' Anna, in the oldest part of town. We were late for mass and had to stand in the back. During the benediction hundreds of women held roses above their heads to catch the blessing. Signora Neri had named both her daughters for the saint, Anna Rita and Margherita. She told me to press my blessed rose in a book. She never mentioned what she'd wished for.

Margherita, her younger daughter, had moved back in with her new husband, Salvatore, who was out of work. They lived in the guest room her mother might have rented. Anna Rita, her older sister, was there most of the time too. She was married and had a house behind the mountain. But she didn't like to stay home alone, so she rode into town with her husband in the morning and ate out of her mother's refrigerator all day.

Signora Neri had a new gas range in her kitchen, but she always cooked on the stove on the porch.

She made her famous couscous, Salvatore's favorite, whenever he brought home some fish. She had learned to make it when she was just twelve; now she was an accomplished master. She steamed the grain and seafood with vegetables in a thick ceramic colander she set on a stew pot. Then she mixed flour and water, rolled a dough snake between her palms, and wrapped it around the base of the colander to trap the steam.

In Sicily everyone is a water gourmet, and every household has its own favorite drinking water. Pharmacies sell Geraci Siculo water in green glass bottles and recommend it to mothers for their babies. The only liquid my cousin Nella would drink was water from her cousin's well in the country. Certain waters are considered miraculous. I have a friend in Santa Margherita whose sick mother sends her to pump water from a well blessed by the Virgin. "There's always a line of people waiting," she said.

Signora Neri's preferred water was the rain that sheeted off the net house roof and sluiced into the cistern below. To bring her water was the least I could do. I strapped her five-gallon plastic jug to my wheeled suitcase trolley, bumped it over the paving stones to the red spigot in the Camparia courtyard, and pumped. I cupped my hand and caught the flow that came up from the foundation. The water of the tonnara tasted like stone.

Signora Neri could do some everyday white magic. She knew the sort of incantations that come in handy around the house. One day Salvatore came home at lunchtime complaining of a headache. His mother-in-law set a chair by the sink. He sat in it with his feet flat on the floor. (If you cross your legs the magic won't work.) She filled a shallow bowl with water, poured a capful of olive oil into it, and balanced it on his head. "If the oil mixes with the water," she said, "the headache is due to *malocchio*," an evil eye curse someone had put on him. "If it doesn't, he just needs two aspirin."

The globs of oil split like multiplying cells and spread into a thin film over the surface of the water. "*È disperso*," Signora Neri judged, and went to flush the cursed water down the toilet. Then she washed her hands, dried them, and placed them on Salvatore's head, her fingers spread above each ear.

"Holy Monday, Holy Tuesday," she said. Then she moved her hands to the front and back of his head. "Holy Wednesday, Holy Thursday," and so on through the Holy Week to "Blessed Easter, that this evil fall from my house!" That was it. Salvatore stood up and snugged on his hat.

"How's the head?" I asked.

"Tutto bene," he said. Everything's fine. He left and the signora went back to washing dishes.

"Somebody doesn't like Salvatore," she said when she heard the front door close.

Rosa of the Cemetery was a good witch, a healer. She brewed delicious cough medicine from honey, herbs, and grain alcohol. (It worked!) She took a shine to me the day I visited the cemetery with Gioacchino to lay flowers on the graves of raises, and she invited me to lunch at her house. She lived in the country with her two sons, four dogs, six cats, a pig, a mynah bird, and wild finches that her son Carlo had caught and caged. Before she cooked our lunch Rosa cooked pasta for the pig. She filled a bucket with spaghetti and gave it to Carlo to carry out to the pig's sty in a tufa pit.

Her specialty was a cure for those who'd lost their voice because of a psychological trauma, such as an earthquake. She told me she had restored speech to a man who hadn't spoken for twenty-three years after the 1968 tremors, when a major quake destroyed my grandparents' hometown. Potions, prayers, the touch of her hands, and her love for Gesù were her only tools, and she never charged for her services.

"Whenever you're in trouble, think of me, invoke my name, and I'll be there to help," she had told me.

Anyone visiting the graveyard would see her sunny face. Rosa Manuguerra was square and meaty, in her sixties, with champagne blond hair in constant need of a touch-up. The cemetery was her social salon, and Sunday, when families came after church, was her busiest day. It was an out-of-town piazza where church ladies in shoes that pinched, helmet hair, and maybe the mink could meet decently, with their husbands in tow, on display. Rosa greeted everyone with open arms, a smile, or a condolence if the loss was recent. The women went to the headstones with buckets and sponges to wash off the dust.

Rosa was a widow herself. When her husband died the town had given her his job because she had his two sons to raise. Although people died on Favignana at the average rate of four a month, the cemetery never expanded beyond its high walls; after two years the skeletons were dug up and placed in an ossuary, where they took up less room. The cemetery had started out as a quarry, but when a stonecutter found a painted crucifix on the rock face he'd just exposed, a sunken church was built on the miraculous site and the graveyard grew up above it. Swallows swooped in and out of the chapel's windowed dome, crapping on the pews.

"Dirty animals," Rosa said. She had to clean it up.

By law, Rosa kept the newly dead overnight in an open casket in a stone storeroom. The law had been enacted after an incident in Tràpani, she said. Having heard a muffled voice in a graveyard, authorities exhumed the body of a recently deceased woman and found fingernail scratches under the coffin lid. Rosa had a gentleman in the storeroom now. "Do you want to see him?" she asked. I didn't.

The dead never complained about the weeds and wildflowers blowing above their graves while Rosa tended to the living instead. People brought their lives to her for a fix. One woman in her twenties with shiny black hair beamed and waved at Rosa on her way out the cemetery gate. "How does it go?" Rosa yelled across the tombstones.

"Everything's fine. He's back."

In Rosa's tiled office, with plastic flowers in vases and the Sacred Heart on the tiled wall behind her desk, I asked her to dictate the prayer to get your boyfriend back:

Saint Anthony, high and mighty
Light an ardent fire in his head
Go to [say his name] and burn his heart and his mind.
If he's seated light a fire
If he's standing cast lightning
He must have no rest
Always of me must he think
And to me come.

"Then say an Our Father," Rosa said.

She taught me other charms that mix prayers to the saints with pagan allusions. To hear news from a distant loved one, say:

Saint Vito of Morreale,
A relative of yours is come to pray
Like a brother or sister of the flesh
Your dogs you must lend me
You must strike [whomever] in the chest,
In the heart, to love me,
In the arms, to hold me
The mouth, to kiss me
The feet, to come near me.

"Then say a Hail Mary and an Our Father," Rosa said.

Il Coppo

When I showed up at the Camparia the next morning, Il Coppo was already loaded in a longboat. Flaxen, the net crested three feet over the gunwale. It humped in the morning sun, fluid curls, waves of woof and warp, charged yet still like a snake poised to strike.

Dry and empty, it weighs two tons. Its outer edges are woven loosely, with the knots in quarter-inch nylon cord about a foot apart. The mesh gradually tightens and the cords thicken toward the center until the nylon cords are an inch thick and the knots touch each other, a carpet thick enough to bear the dying bluefin, tight enough to hold water.

I asked Clemente when the mattanza would be. Maybe tomorrow, he said. But days passed, and Il Coppo remained in port, brooding like a burial mound. It filled the twenty-foot boat.

I spent my days talking to the men while they repaired nets in the sunny courtyard, or cleaned the cavernous net house, now empty, or sanded new oaken parts for old anchors. They worked in slickers and thick sweaters in a misty light rain and grilled fish for second breakfast in the boathouse when it poured.

One afternoon, when Vito came back from the one-thirty check of the trap, he said, "Come with us when we set Il Coppo."

"Tomorrow?"

"Probably. Don't say anything."

When work was done for the day we went to the Two Columns for a drink.

Vito said the Parodis had hired lawyers to fight Castiglione. Rumor was that the Parodis wanted the tonnara back at the end of the season—the end of Castiglione's nine-year lease—just as they had given it to him, with the old wooden boats ready to go back in the water. Castiglione had substituted his own brand-new metal boats, and the Parodis feared he'd take them back with him to his own tonnara at Bonagia, across the strait, the only other tonnara left in Sicily, leaving the Parodis with a fleet that was not seaworthy. "It's not right," Vito said.

Little Vincenzino, the octogenarian ex-tonnaroto, said, "Castiglione has no intention of catching a lot of tuna this year. If he catches six or seven hundred, enough to pay the fishermen, he'll take the rest in subsidies. The rest of the money he takes from the government and puts in his mattress to sleep on."

Vito said it would be good for the tonnaroti to form a cooperative to lease the tonnara from the Parodis for five years at a time. "But it takes the goodwill of the rais, and he is always speaking poorly of the men," Vito said.

The next morning I arrived at the boat dock at seven. Ships' flags snapped, the wind blew up whitecaps, the ferry rolled drunkenly at its mooring. Setting Il Coppo was a piece of cake on a calm day, but difficult and dangerous when there was a sea. It was late in tuna season. When the rais decided to try to set Il Coppo the men looked grim. I was not invited aboard. Vito shrugged and spread his arms, as if to say, "It is not I who commands." I resigned myself to staying ashore with all the other journalists now milling around and hoping to be invited aboard.

Close to mattanza time the tonnara becomes a media circus. Some reporters were on assignment from Italian newspapers and European magazines, some were freelance journalists, some were amateur photographers, khaki vest pockets stuffed with film, hoping to get published with a dramatic shot of men killing tuna. We were all cooling our heels dockside, but a Japanese television film crew was aboard the *Nino*, Castiglione's big workboat, which would tow the tonnara cortege to sea.

A Japanese television network had sent two full crews to Favignana for ten days to film a story on the mattanza. They were getting the royal treatment from Castiglione. One crew was based in Paris, I learned. The other had come from Tokyo. They brought with them a popular

Japanese actress, small and soft-spoken, with seemingly unlimited access to the rais in his dark office.

She and the rais were often in front of the picturesque Camparia gate surrounded by lights, cameras, and microphone booms. On camera he would smile and say something charming for her. This morning the actress held an ice bag to her cheeks. Matteo Campo asked her if she had a toothache.

"No," said an interpreter. "She does this to make her skin more pallid." She was getting on my nerves.

Only the Japanese were invited on the boats. Close to tears, I watched Il Coppo towed to sea without me. Then I thought, *What is money for if not a time like this?* I jogged to the fishermen's port and looked for Sebastiano, the old fisherman who had taken me to see *u cruciatu*. He wasn't there, but a friend of his was.

We were almost out of port when a freelance writer from Messina and a photographer from Palermo waved us down from the ferry dock. The rest of the pack ran up behind them. Messina agreed to take three of them. The others hired other boats, a small bonanza for retired fishermen. I asked Messina, an old tonnaroto, to keep the sun at our backs when we got to the trap. It felt good to be on the water; the sea spray revived my spirits. We spent an hour bucking choppy waves near the trap until the rais aborted the effort.

The journalist and the photographer were disappointed. They were serious amateurs who had come months earlier to get permission to go to sea with the ciurma, and this was Friday. If Il Coppo wasn't set, the mattanza couldn't happen this weekend and they would have to go back to their day jobs and miss it.

Next morning I was back at the dock at six, the first journalist there. The sun was bright, the sea was flat, Santa Caterina was lit up like topaz. While I stood taking pictures, there was a sudden, silent rush of activity. Capoguardia Messina, the oldest man in the ciurma, walked up behind me and quietly told me to get on the boat with the net and hide, and to play dumb if any journalists asked questions. There was even talk that there might be a secret mattanza right then and there.

I stepped into the boat, picked my way across the mounded net, and sat on the seaward side where I would not be seen from the dock. I pressed my spine against Il Coppo, leaned back, and looked at the sky. I was the only journalist they brought. I wondered if the rais knew I was aboard.

We cast off. I clasped my hands behind my head and watched the shoreline recede, the green dome of the church shrink, the point of Bue Marino come into view. When we arrived at the Chamber of Death, the rais gave me a hand into his musciara.

The men started unloading Il Coppo right away. Sailboats and yachts gathered to watch. Everybody had heard the rumor that there might be a mattanza. The rais yelled to the nearest boat captain that there would be no killing that day, but they hung around anyway.

The rais relaxed; the men knew what to do. "Today I'm a guest," he said, and gestured to the men in a square of boats unfurling the net like a tapestry. "This is the easiest part." Where the white net was hung the lapis sea lightened to powder blue until the whole square was glowing like a gem lit from within. The net's belly sank, the mesh stretched taut, slanted down into the water, and disappeared. It hung there like an empty womb.

I'd never seen the rais so relaxed. The men in the musciara threw fish bait to a pair of seagulls hovering overhead. Young Antonio threw a viole to the gull on the right, and an older crewman chastised him. "No! Don't feed him. Feed the mother!" Somehow, instinctively, he knew what Joseph Campbell knew: "Woman is what it is all about—the giving of birth and the giving of nourishment."

The rais opened a cupboard and pulled out his spyglass, a metal drum the size of a wastebasket with a glass bottom. Kneeling, he leaned over the gunwale, pressed the glass into the water, and peered through it. He let me look. There was a great blue emptiness in the Chamber of Death.

The men knew I'd been upset the day before when they'd left me behind. Today they asked me if I was all right. "We pulled a few strings to get you aboard," Capoguardia Messina said.

The Japanese film crew loomed up in the distance, on a tour boat they had hired. The cameras were rolling. The actress cupped her hands and yelled, most politely, "Oggi mattanza?"

"No mattanza today," the rais answered, smiling. "If there were I would have called you." He motioned them closer. The seagulls hovered tightly above his raised hand, thinking he would throw them something. The Japanese, apparently misunderstanding the gesture, drifted away. The rais stretched out on the bench and put his head on his arm.

"Leave me here," he said. His men knew he had to be kidding. "On land there's the telephone that rings every minute, you can't even eat." It was always a journalist or a travel agent wanting to know when the mattanza would take place. The rais closed his eyes, and I started to shoot. He lifted his head. "You take pictures of me while I'm sleeping?"

"Yes. Last year I got quite a few." He lay down again. "Send me some of those pictures of me sleeping so I can show my wife. She always says I do nothing all day and that will give her fuel." He closed his eyes.

"Guess what the Japanese said to me. They said they had been afraid of me. But now they know me, they said. They didn't think I could be like this. Are you afraid of me?"

"I used to be. And sometimes I still am," I said.

"Why?"

"When you get angry, Salvatore . . . "

He laughed. "When I'm angry I'm not angry at you. Why are you afraid of me?"

"When the pot is boiling you can burn yourself if you get too close." An old saying I just made up. He laughed. In an hour the men had finished their work. Someone threw us a line, Rosario slipped the noose over the bow, and we were towed home. I took out my brown bag and offered the rais some cheese and bread. He wasn't hungry, he'd just eaten a candy.

"Two," I corrected him. I'd been watching him closely.

"Talia!" he said to his crew in Sicilian. "Look! She counts the number of candies I eat."

I tried changing the subject. "You were like Saint Francis today, calling those seagulls. It's going to be a beautiful picture."

"I wasn't calling the seagulls, I was calling the Japanese."

"You have the magical powers of the rais."

"You read too many books," he said. "What I would like of this magic is four hundred big tuna." He said he would go home now and take a nice two-hour nap. "What do you think of that?"

"You're the boss."

"I'm not the boss of anything," he said. He moved from the seat down to the floor of the boat at my feet and took my left hand. He twisted the cameo ring I wore on the third finger. "You should marry," he said.

"I'd be an awful wife. I like my life as it is. I like being able to come to Favignana and spend a few months with the tonnaroti."

"There are a couple of tonnaroti looking for wives." He was serious. "But they are too young for you."

The sea was magic in the spring; I was sorry to see land again so soon. A small crowd waited at the dock; people stared at me as I debarked. A man on a bike stopped me.

"Was it beautiful?" He was jealous.

"Yes. Everything they do is beautiful."

"Describe it to me." He felt cheated. "So many years I've been coming here. . . . "

"It is not I who command," I said softly. "I was invited."

"I understand," he said.

When I collected my bike and walked toward the piazza Capoguardia Messina came up to me. "Now, then, are you satisfied?"

The sight of Il Coppo sinking into the sea had been the tonnarotis' gift to me. I thanked him, sincerely, from my heart and with the light of my eyes, and he walked away happy.

The Bell

The Camparia's little bell, forged in the Florios' foundry and rung only on the eve of the first mattanza, had not been heard for a year. Life happened while we waited for it to ring.

I was still stalking Vincenzo Sercia, the old First Voice, with my tape recorder, hoping he would sing the cialome for me. Sercia spurned the piazza's bars and men's clubs and socialized instead with the retired fishermen who congregated in one of the tufa sheds overlooking the port. The sheds, former servant quarters and carriage houses, formed a square near the Florio palace.

Sercia and company spent their days in the dark room. From its doorway there was a panoramic view of the Camparia, the cannery, and the seawall. They wore berets, turtlenecks, beaked caps, and tweed suitcoats. They sat in the penumbra on stools and rough wooden benches under the arched, vaulted ceiling and talked. They watched the people who stepped off the hydrofoil, the trucks that boarded the ferry, the fishermen hawking their catch, the passing boats, and the progress at the Camparia. When they were in, the door was open.

I knew none of them, but they knew me and found a stool for me when I came to visit. After a minute my eyes adjusted to the dark. Caned fish traps and fragments of net hung from nails in the walls. Sercia opened the fridge and offered me a cold beer. He had been telling jokes again. His friends had heard all of them, but they asked for them by name. Here it was the art of telling that was important. The men listened, rapt, and laughed on cue.

They asked Sercia to sing for me. He wouldn't, but pushed forward a friend who would. Bertuzzo was eighty-four, his face like a dried apple, with a blue beret pushed over one eyebrow. Sercia sent him home to get his guitar. He came back with a black vinyl case, unzipped it, and took out a fifty-year-old guitar whose strings hadn't been changed in years. Its black-lacquered body was inlaid with a scene in mother-of-pearl—the silhouette of two men serenading a signorina on her balcony under a full moon.

Bertuzzo strummed a sour chord and began to sing in a frail, high-pitched voice. Love songs from his youth, Sicilian songs forgotten by all but the old. His buddies sang along softly, faking the verses they'd forgotten. Pasquale

Mercurio, a handsome seventy-four-year-old, kept the beat with mouth music, a huffing *v-v-v-v* sound he made using his lungs as a bellows. Sercia hushed people who stepped into the dark to listen. It was like stepping into a time before radio and television, when people still knew how to amuse themselves.

When Bertuzzo put down his guitar, they told stories about the old tavernas. How many there were! They closed their eyes and counted them up, and told me where they used to be. They told me how you could go to a taverna, choose an octopus from a tank, have it boiled, cut it up in bite-sized pieces, and squeeze a lemon on it. You could get wine, watered-down wine, that made some people rich. You could play cards, and bet and win or lose and make your wife mad. There was always a guitar hanging on the wall, and someone who would take it off its nail and play it. A boy could come in and sit on his grandfather's lap, and when he got his kiss he'd feel the scratchy beard on his cheek and smell the wine on his grandfather's breath. There were no bars or pizzerias when they were young, just the dark tavernas where fishermen and quarrymen met.

When they were young they had been tonnaroti. I asked what the tonnara used to be like.

"The Parodis used to give a tuna every year to the people of Favignana, a tuna from the first mattanza of the year. They used to cut it up and give half a kilo to every family in town," Sercia said. "Now we don't even see the color of the tuna." He stood in the doorway and looked at the cannery's cold smokestacks. "When they were cooking the tuna . . . ," he said, and trailed off into a reverie. "The aroma . . . it made your mouth water."

We had come to May 22, the day the first mattanza was held last year, but still the bell was silent. The tonnaroti awaited word from the rais, and the rais was waiting for something—some said the Japanese freezer ship, some said Castiglione's important government guests. Maybe he was just waiting for more fish.

I had just left the library and gone to the Camparia with a book written by Raimondo Sarà, *Tonni e tonnare* (Tuna and Tonnaras), about the traditions of the Favignana tonnara. The author was a retired marine biologist, an expert on the bluefin tuna, with fifty years of field research behind him and a long stint as head of a regional government institute on fish and Sicily's fishing economy.

The rais was in his office, within earshot of the Camparia gate where I was entertaining Aiello, the gatekeeper, with the story of my one and only marriage proposal. It had come in my grandparents' hometown, Santa Margherita Belice, near the end of a two-month stay with my cousin Nella.

During supper, long after visiting hours, the doorbell rang. I opened the door to find a short man in his seventies, hat in hand, whom I'd met once on the bus from Palermo. He brought this message:

"Signorina, if you do not wish to leave Sicily, there is a man here who would like to marry you. He is rich. He owns many acres of prickly pears."

Apparently my suitor had seen me walking with Nella in the Most Holy Crucifix religious procession, the classic way to find a spouse in a provincial town. Nella figured out who he was—a well-to-do farmer in his seventies who lived with his sisters, and whom she had earlier turned down. She bent over double, laughing. "He doesn't even speak Italian," Nella said, "only dialect. You wouldn't even be able to understand him!"

I thanked the man politely, but declined.

"Think about it," he said. "You don't have to decide right away."

The rais called me into his office. A distinguished man with a cane sat before his desk. "Why don't you get married and quit writing books?" the rais asked.

"Can you see me as a housewife?" He must have thought I meant that I didn't know how to run a household.

"You've got to get into training, be prepared," he said.

In this society, if a woman doesn't have a child, preferably male, she is not complete. Break a link in the reproductive cycle and you're pitied in Sicily. I'd had enough. "You think that if a woman doesn't have children she is not a lady," I said.

The man with the cane spoke up. "That is not what he said."

I looked the rais in the eyes. "He said it in the musciara once. You spoke of the Japanese actress. You said she had two children, grown children, like yours. 'She's a signora, a lady,' you said. I have no children. What does that make me?"

He broke into a toothy grin. "You're jealous!"

Years later I cringed when I realized that *signora* can mean "lady," or simply "Mrs.," a married woman. But just then Aiello called to me, saving me from further embarrassment.

Antonio, a towboat captain, was going out to the trap, he said, and he wanted me to see a live tuna that had just got its fins stuck inside the trap. I placed the library book on the rais's desk and asked him with a look for permission to go. He waved me out of his office without a word.

We motored out to sea with Girolamo, the diver, who was to cut the tuna free. The crew seemed content. Girolamo suited up aboard, but we arrived too late: The tuna had suffocated. Girolamo cut the corpse free and tied a rope to its tail; with great effort, the men pulled it aboard. I leaned over the gunwale to watch it come out of the water—silvery blue, marbled, metallic, a good four hundred pounds. The men smiled, happy to think that more like this could be swimming in the trap.

When Alberto thrust a gaff into the fish I heard a ringing thud, like a ripe watermelon when you slap it. The sound reverberated in my ears, and I had a sudden visceral understanding of what the mattanza was for the tonnaroti. The living flesh, just dead, or now dying, in their hands. Alberto caught me gaping at the fish over the top of my camera. "Pesce fresco," he said with pride. Fresh fish. "That's why it made that sound."

Back on land the word was mattanza tomorrow, but nobody would believe it until the bell rang. I found Gioacchino repairing a net. He was feeling low because Castiglione would not allow tourists to watch the mattanza. They were an insurance liability, but Gioacchino had invited friends who were flying down from Milan to watch. He was despondent.

"Everything is more beautiful after the mattanza. The mattanza is a heavy burden," he said.

In two days Gioacchino would turn fifty-three.

"I used to walk in the Chamber of Death after the mattanza and carry the small tuna, still living, in my arms to the longboat." His arms cradled an invisible baby, like the Madonna of the Tonnaroti cradling the tuna with the turquoise eye. "With the tuna I have felt all the emotions. All of them."

The rais summoned me to his office a second time and nodded toward the man with the cane. "The gentleman wants to ask you a question."

"Signorina," the gentleman said, "I'd like to ask you why you brought that book to the rais's office." He'd seen my copy of *Tonni e tonnare*.

"I'm writing about the tonnara, and I use the book in my research. Why do you ask?"

"Because I wrote it."

"You are Sarà?"

For years in the spring he had roamed the Mediterranean visiting tuna traps from Spain to Turkey, measuring water temperature and salinity, interviewing tonnaroti, tagging, weighing, and dissecting bluefins. Now he was retired and wrote books. *Tonni e tonnare* was the first he wrote for the general public; he said he was writing another, about the origins of the tuna hunt, before Christ. The name for Canaan, the promised land, he said, derives from the Hebrew word for tuna, "*thun*, and it's written like this." He walked over to the Camparia's wooden gate. With his finger, he traced a hieroglyph, a lazy diamond with a line for a tail fin. It looked just like the scar on Clemente's shoulder. Clemente emerged from the "taverna" storeroom in a sleeveless T-shirt. I asked him to turn around and show Sarà the fish etched into his skin. "Look at this one!" I said.

Sarà smiled at Clemente. "Yes, I know this one."

Sarà knew the most arcane stories about the tonnara. At one Mediterranean tonnara the fishermen do not pray the litany, he said. Instead they keep an urn full of saints' names at the courtyard's entrance and choose the patron saint-of-the-day at random every morning of the tuna season. He knew of a rais who, in bad years, ordered a man to be dressed in white robes and dunked at the bocca di nassa to frighten the evil spirits that kept the tuna from entering. He told me about the time thirty years before when the tuna would not enter the Favignana trap, so the rais took the town prostitute to urinate over the trap entrance. With this ritual, and a little luck, he hoped the tuna would enter.

ꝏ

By the end of the day Castiglione had arrived and agreed to let tourists watch the mattanza if they signed a release form. The rais sent Gioacchino to the tobacco shop for paper, and Castiglione dictated the conditions. I wanted to be the first to sign, but the rais said this was not necessary.

"You are a tonnarota, a member of the crew."

At four-thirty the ciurma straggled into the courtyard and watched as Castiglione gripped the rope that hung from the bell and yanked it. It rang like a fire alarm.

"Forever be praised the name of Jesus!" Occhiuzzi shouted.

"Gesù!" the men replied. "Gesù! Gesù!"

Domani mattanza.

La Mattanza

May 23. The sirocco blew, the sea was rough, not a good day for a killing, but the rais seemed to want to have done with it. After having my cappuccino at the fishermen's Bar del Corso in Piazza Europa, I stepped out into the street. Clemente and Angelo wheeled down Via Roma, Angelo on his antique motorbike, followed by his father on a bicycle, followed by a van full of Japanese videographers, microphones stuck out the window, cameras rolling. I left and walked over to the Camparia.

The Japanese actress was shivering, so the rais gave her his heavy orange rain slicker to wear. She rolled up the sleeves four times. I boarded a boat loaded with tourists, and we headed out to sea in beautiful slow procession.

At the trap the sun was hot and the sounds were soft. The palms bent in the wind. The ciurma killed 290 tuna, all small—four- and five-year-olds, no more than 60 pounds each. It was an anticlimax for the film crew. Maybe Clemente was right, maybe the big ones had passed.

When it was over Clemente dived into the death chamber and swam across to Margo, who had brought their child. He climbed aboard, still dripping from the bloodbath, and took Callista in his arms.

As we were towed back to port, I asked a Japanese cameraman, in French, our common language, why they had brought an actress if they were making a documentary.

"It's a typical Japanese thing," he said. "The idea is to have the actress develop a sort of relationship with the rais on film, to better engage the emotions of Japanese viewers."

Up until this morning's disappointing catch the film was supposed to be a ninety-minute special, a major production, he said. "But now I can't imagine it running more than fifteen minutes."

The film crews had been here ten days and were leaving tomorrow for Paris and Tokyo; their budget was spent, and there was no chance they could stay and film the next mattanza, which would probably take place in another ten days. For them it was over. But I would stay, for at least one more mattanza, maybe two. "The fishermen speak highly of you," the cameraman said.

Back on land the tonnaroti told me that every year they have one mattanza of all small fish, but it's usually the season's last. Despite the disappointment, the tonnaroti seemed calmer, and there would be more mattanzas this year. All fishermen are gamblers, and gamblers are optimists.

The tonnaroti asked me, "Did you like it?" I lied, a white lie, and said yes to everyone except Clemente. During the mattanza I had capped my camera lens when I saw how small the tuna were. It seemed a crime to take such little ones when I remembered the behemoths of years past.

That afternoon I went with the rais to check the trap for new arrivals. The Japanese were packing up their equipment, the Italian journalists were off to Palermo and Messina or to catch flights to Rome and Milan, but I was still here, in a boat with the rais, under the warm spring sun.

Rosario chanted the prayer. The other towboat pulled up beside the musciara with Gioacchino, bare-chested, sprawled on the bow. Michele Pattì, an oarsman, said Gioacchino was beginning his summer regime "to get good and tan before the season begins, so when the tourists get here he is in shape." The men took their turns counting tuna under the tarp. Vito cut bait on a pink terra cotta tile he kept stowed in a cubbyhole and then dropped his line in the water.

"I'm going to sleep now," the rais said, and curled up in the shelf of the musciara. "If you take my picture I'll kill you." I was glad he did not treat me as a guest. Just the same, I capped my lens. He closed his eyes, then peered at me from under his eyelids.

"You're too timid," he said, his head cradled in the crook of his arm, curled up like a child, napping in the sun, at peace.

This was my chance. "I would like permission to board the longboat during the mattanza to record the songs," I said, boldly.

"You can get on any boat you want."

The rais took a turn under the tarp, counting tuna. I watched with him. The nets stretched down and vanished. Sunbeams splayed through the mesh. It bulged with the eastern current, taut as the skin on a pregnant belly. The tonnaroti watched for the fish, like lovers, with adoration in their eyes, eager for even a glimpse. Perhaps they saw themselves mirrored in the bluefin.

I marveled that so soon after a killing more tuna would enter the trap. "You would think that they somehow would know that a massacre just took place here," I said.

"The sea forgets," said Michele Pattì. The currents had washed the sea clean of blood. The south wind had calmed. Three of the crew were napping. The boat rocked on the breast of the sea. A watcher in the next boat shouted that a fish had entered the Great Room.

I asked the rais, "Does the government pay Castiglione?"

"Who told you that? Castiglione has been waiting for money from the government since 1989, when he had a new vascello built." The vascello was the longboat from which the tonnaroti killed. The government had promised to pay him half the cost of new materials, the rais said, but "the region hasn't paid him yet."

"Why don't the tonnaroti form a company and lease the tonnara, as a group?"

"Who would buy the tuna?" the rais asked.

"Castiglione," I said.

"You think so?"

"The Japanese then," I said.

"Who would supply the refrigeration?"

I had no answer, but I thought to myself two things. One, where there's a will there's a way. And two, no doubt he's already thought about this thoroughly.

"Who will loan us the money?" the rais asked me.

"A bank. The region. Money out of your own pockets," I said. The thought hung in the air, then the rais changed the subject.

"The Japanese are giving a party tonight, but I don't think I'll go," he said.

"The actress will be disappointed," I said.

"She's married. She has two children."

"She will still be disappointed."

"I'm married. I have two children."

Rosario told me that when the Japanese came to take their official leave of the rais, the actress had said, in English, "I like you."

" 'Like.' What does that mean?" the rais asked me.

"Tu mi piaci," I answered. You please her. The men hooted, and the rais turned red.

"I told her I didn't like her," he said.

"I don't believe it."

"You don't believe me?"

"No."

I told the rais I would go to Santa Margherita Belice for a few days between mattanzas.

"Do you have a boyfriend there?" he asked.

"No."

"You should get married."

"Why?"

"Make a family. It's good."

This conversation was going nowhere. I reached under my seat and drew out my brown-bag lunch, thick-crusted yellow bread and provola cheese, but I didn't have my knife. "Does anyone have a knife?" I asked. Half the crew was asleep.

"Tiene, Teresa." The rais sat up and handed me his pocketknife. When I finished I wiped the blade on my jeans and closed it. The rais was dozing, but when I dropped the knife into his open palm his fingers snapped closed on it like the bite of a bluefin.

He took me to sea often. He said I was good luck. I said I never had any effect on the luck of Piero, the fisherman in Mondello.

"If you are bad luck, we will leave you with San Pietro," the rais said. Just then Rosario caught a couple of viole.

"Maybe my luck is changing," I said.

Rosario took the rainbow fish off his hook. "This is not luck," he said.

"What is it then?"

"Luck is when the fish enter."

In a few minutes ten or fifteen huge tuna swam into the Great Room. The man under the tarp saw them and called out in a muffled voice, his face pressed to the glass. "Sono entrati!" The rais turned to me and said, "Tomorrow, Teresa, you must be here." Then he booted the crewman from under the tarp and took his place. The rais said he saw twelve or thirteen fish. Rosario, looking through the glass-bottomed bucket, said he saw twenty or twenty-one. The men in *Jupiter* said they saw twenty or thirty. The rais looked for them in vain.

"Where are they?" he bellowed from under the tarp. He

was angry, yelling that the men in *Jupiter* hadn't really seen thirty fish. "They must be hiding," he shouted.

ꝏ

At the second mattanza, on the second of June, they took eighty-seven tuna, of small to medium size. Today the third mattanza would take place.

I had dreamed the tuna dream again. I was one of them. I swam with giants in a cottony silence, coasting the walls of the Chamber of Death, all eye and muscle, the net's soft bent diamonds a peripheral blur. Then the net contracted, then again, and again, a throb, like some muscle convulsing, squeezing us together as we rose slowly and unwillingly toward the singing. When I hit the bright light, I woke up.

It was late. The square of color that was my clock had slipped down the wall. It was already half past red when I threw off the covers. I skipped breakfast and biked into town like a streak.

A crowd had gathered long before seven in front of the Camparia. The press was back in force. Bill Allard, the *National Geographic* photographer, was there, and Giò Martorana from Bagheria was shooting black-and-white for his photo essay book, *Tonnara*. Luciano Bovina, a documentary filmmaker from Bologna, had brought a still photographer and a writer with him. The *Giornale di Sicilia* was there, and a freelance photographer from northern Italy, and Antonio Noto, the bookseller who was the best photographer on the island. And there was a polyglot television crew: The producer spoke American English, her boss spoke Greek and English, and the crew conversed in German. Hundreds of tourists milled around.

In fifteen minutes we were at the trap. Retired fishermen made a few thousand lire ferrying latecomers to the scene. They pushed their cargo over the gunwale into our already overcrowded longboat. The sun beat down. I staked out the best vantage point over the Chamber of Death for Bill Allard and me. He was paying me to write some caption material. The unruly adolescents aboard scooped up seawater in hats and plastic bags and threw it around. The photographers roared and covered their cameras with their shirts. Giò told the kids to have some respect.

"What did you come here for anyway?"

"To have fun."

Giò told them to behave. The wait went on for five hours. I had brought a book, *On the Road*, to pass the time. It felt strange to be racing through the American Midwest on the back of a flatbed truck, then to look up and watch seagulls and feel the weave of an ancient net under my seat. It didn't seem right, like knitting in church, so I closed the book.

The currents were running the wrong way; the shoal could not be chased into the Chamber of Death. The chaser net, the ancera, would not unfurl correctly. The tuna they did scare into the Chamber of Death turned around and swam back into the bastardella.

The rais aborted the mattanza, almost unheard of. It was *brutta figura* too, because Castiglione had brought guests. Six hundred people were disappointed, including the rais and his men. Shortly after noon we were towed back to land, where everyone rushed for something to eat, a cold beer, and the bathroom.

The ciurma reassembled back at the Camparia gate and waited for the word to break for lunch. The rais strode back to his office in a funk and waved them off without lifting his eyes. At the boats, amid the stink and roar of small motors, Clemente yelled to his son Angelo.

"Be sure to be here tonight at five, with your mattanza clothes on, if you want to make that money."

It was easy money, Clemente told me. The polyglot television crew couldn't afford to wait around until the next mattanza and so had offered 100,000 lire, three days' pay, to every member of the ciurma who showed up for the filming of a faked mattanza.

They all showed up, of course. All they had to do was pretend to pull up Il Coppo and sing. The rais had agreed to let them use the vascello as a prop if it stayed in port. While they were filming, the huge white ferry pulled up to its anchorage in the harbor in the background. The tonnaroti roared with laughter, for it showed the trap was fake and not at sea. They sang their songs as they pulled up a limp scrap of net. They shouted as if they had tuna coming up in it and gave the cameramen their money's worth. Little old Momino had the bright idea to leave the silly net and go stomp on the floorboards of the boat to imitate the tuna's thunderous thrashing. The director yelled to the captain of the towboat: "Three more meters this way! Two meters that way!"

That night Bill Allard stood with Clemente at the Two Columns bar.

"How long will the tonnara last?" Bill asked him. I translated, and Bill wrote the answer on a cocktail napkin.

"There will be a tonnara as long as there are fish," Clemente said.

On June 8 the wind changed to *bonaccia*, and the sea was still. Tomorrow mattanza, people said.

Next day the crowd was thin at the Camparia. It was the middle of the week, and most people were at work, the novelty had worn off, and the off-islanders were few. Most of the photographers, disillusioned with the small, few fish and the aborted mattanza of a few days before, had gone away. This could be the last mattanza of the year, and it would almost certainly be the last I'd see that year. Most of the gathered people were Favignanesi, friends and family of the tonnaroti. It would be an intimate killing.

I was about to board the tourist vascello when Gioacchino told me to get in his boat; they would take me to watch the fish pass into the Chamber of Death. After the aborted mattanza Angelo had told me he had seen a giant white tuna sprinting through the gate into the Chamber of Death, but then the frightened tuna had turned and sprinted back into the bastardello. "Maybe you will see it," he said.

The marine caravan formed in the shallow green waters under Monte Santa Caterina. Men threw lines boat to boat, the rais shouted commands, the people on the vascello waved to their friends on the dock. The tonnaroti tried to work around the landlubbers, asking them politely to move, politely to sit down, politely to get out of their way.

Then they stowed their gaffs, took their stations, and stood tall at rudders and bows. The boats moved out slowly, towed in two long lines. Once out of port the captain of the rais's musciara shouted, "Una salve regina alla Madonna di Tràpani!"

The tonnaroti removed their hats. Anyone who doesn't take off his hat for the litany gets his cap snatched from his head and thrown in the sea, young or old. We pulled up behind Saint Peter's back while the other boats formed the square above the Chamber of Death.

Gioacchino transferred me to the *Saturno*, Momino's boat, which would stand watch above the flowered gate of the Camera della Morte. As soon as we tied up to the bastardella cable the *Saturno*'s crew rushed to cover the benches with a tarp, untied the ropes that held the gate closed, and lowered it into the sea. Behind us men in two small boats rowed the length of the trap east to west with the well-hung ancera stretched between them, driving the trapped tuna toward the Chamber of Death. I watched with Angelo under the tarp.

The sea was thick with floating motes. I saw five bluefin, murky black diamonds, coast casually through the gate. Angelo yelled, and the five men in the *Saturn* heaved with all their might to pull up the heavy wet gate through a hundred feet of water. But the five tuna swam back over the transom before the gate was completely closed. They streaked under our noses.

"When they fold their fins in they go like mad!" Angelo said. Angelo could hardly contain himself. "There's a big one in here. There's a white one in here. I saw it up close." He hoped the big one would come to him. Angelo is an arringatore and works the *speta*, the shortest gaff, at the center of his killing team, the same position his father has always worked.

"I feel sorry for the small ones," he said. "They come to an awful end."

Off in the distance we heard the crowd cheer whenever the gatemen pulled, their backs arched, gritting their teeth, windmilling their arms to pull up the gate before the fish could turn around. They could pull it up in under thirty seconds, but it usually wasn't fast enough. Their colleagues shouted encouragement, but the wind blew their voices to the base of the mountain.

It took forty-five minutes to herd the fish into the last room of the trap. The *Saturno* took its place in the square. To my delight, Gioacchino ferried me to the vascello where the men would kill and the tuna would die. This time I would see the sacrifice from the altar.

I sat in the stern, feeling like a trespasser on sacred ground and trying to make myself as small as possible. But it was the tonnaroti themselves who had brought me here. None of them mentioned the rais or asked if I had his permission when they shuttled me from boat to boat.

Berto prepared three lengths of rope. For every ten fish killed he would tie a knot. Gioacchino told him to get the count right because Castiglione's numbers did not always agree with the fishermen's, and their small bonuses depended on the number they killed.

The rais sent the diver down to check for holes in the net or the presence of swordfish. When Girolamo surfaced he handed the rais his air tanks and was pulled aboard the musciara. The crew rowed the small gray boat around the square and picked up the tonnaroti stationed on other vessels. Twenty-five men stepped aboard the smallest, oldest boat in the fleet. They looked like refugees. When they boarded the vascello they slowly took up their places with an air of concentration. Rosario tied the musciara to the rais's cable, which stretched diagonally across the Chamber of Death; it floated in the middle of the square.

The rais squinted into the sun and pulled on his raincoat. For the first time I would see the rais's face as he directed the mattanza. He did not perform the old rite of shouting the name of the hunter to the hunted. An old rais would have said:

"Arise! Arise in the name of the Lord! Sugnu Rais Mercurio e vi saluto!" (I am Rais Mercurio, and I salute you!)

Salvatore Spataro, in this last decade of the millennium, simply raised his arms, the silent command to start pulling. The men lined up, the sun on their backs. Nonchalantly they pulled Il Coppo aboard. They did not sing.

They leaned back in waves, a ripple of shoulders. They pulled unevenly, a strange quiet about them. I did not watch the tuna rise to the surface but leaned against a mast and watched the men's backs. Clemente was in his lucky rag of a shirt with a thunderbolt on his chest. Gioacchino wore a new red T-shirt and the shredded shorts he'd worn through many mattanzas.

I saw the tuna as the men saw them, face on, flesh ringing in the gaffs, each one a muscled weight out of water. They weighed several hundred pounds each, a few were massive. Clemente and Gioacchino killed together. They lifted a giant; it whipped its tail in spasms and swam in the air at the end of their gaffs. I read no emotion in the unblinking black eye, but the fish had turned coppery gold. One after the other the tuna gasped and thrashed, then, stunned and defeated, slid into the hold. Thrashing, they bled and died at my feet.

The tonnaroti say if a man falls backward into them he will come to no harm because man's touch sends the tuna into a trance. "S' inniu," they say in Sicilian. He went away; the spirit has left the body.

They killed 167 fish that day, but not the white one. It had remained in the bastardella with a group of ten others that were not ready to cross the flowered threshold. The tonnaroti threw buckets of crushed ice on the silver bodies and then covered them with a tarp. The vascello full of bluefin was towed to Tràpani; the islanders never saw their colors.

Later I met Antonio Casablanca at a table outside the Two Columns. He pulled his blood-splattered shirt over his head and gave it to me. Then he slumped in his chair and stared over his Campari.

La Salpata

There was to have been another mattanza on June 11, but during the night a great ship ripped up part of La Costa, the eastern barrier net.

"An oil tanker, it was!" said Angelo. He was excitable. "Big as Formica, it was!" This came as a blow at the end of a bad season of meager catches of mostly small fish.

There was little for the men to do but mope about the hard work to come, now complicated by the accident. Half the ciurma had the day off, the rais had gone to check the trap, and the few men left on land were in the cool, cavernous net house getting ready to fill it with the nets as they were brought back to land.

I sat on a dusty palette in an alcove. The pillar at my back bore the black imprint of the palms of two large hands, an Arab talisman to ward off bad luck. Beside the palm prints a year was brushed in crude black strokes, "1983," and under the year was a number, "854 tonni," a brief summing up of a bad season. On the wall of the empty alcove, like an icon in a side chapel, was the image of a sailing ship. These net house graffiti, made by hunters, lay hidden, like the paintings in the Levanzo cave; they were visible only in the spring, when the nets were in the sea. Soon they would be hidden again.

To amuse the workers and take my mind off leaving, I described the taverna I would own. It would be in Via Roma, close to the piazza, with rough tile floors and booths with wooden benches, dimly lit so the dirt would not show. A guitar and a couple of tambourines would hang from pegs. I would offer a 20 percent discount for pensioners. Now I needed a name for the place. There was already a restaurant called U Rais, and another called Musciara. Capiletto suggested Da Teresa, but I wanted a tonnara theme. Matteo Campo parked his broom and sat down to think with his face in his hands.

"We can call it La Camparia!" He even had a slogan that rhymed in Sicilian: "La Camparia—cu mangia e cu talia." Some eat and some look. One by one the men left their work and sat with me and Matteo. First they brainstormed the menu. Boiled octopus was a given. "And *fave bollite*," Matteo said. Fava beans, boiled potatoes, roasted artichokes, hard-boiled eggs, and, once a week, pasta with garlic. "And you can get rich on watered wine," Giuseppe said. He knew just the man to advise me.

He promised that the tonnaroti would bring potted plants to the grand opening, and the rest nodded in assent. We conjured this taverna in the cool void under the pointed arches until Matteo had an attack of work ethic and stood up slowly.

"You can hire me to sweep," he said. So I had potted palms and a janitor, and that was a start. But the scrape of his broom was the sound of desperation, and my daydreams eddied up with the fine dust and fish scales.

The end of the mattanza season was closing in by the second week of June, after which the *salpata*, the weighing of anchors, would begin. When the Russian oil tanker's propeller tore up eight hundred meters of La Costa, it had mangled the steel cable and dragged the precious old anchors halfway to Marsala. The rais had to leave the mess in the water for inspection by the insurance company and the police, but he needed to pull up a few anchors before they were lost altogether. The salpata began early that year.

I was in a boat whose four men were to pull up ten anchors with a winch. The long swells towered over our heads and rolled under us like lumbering giants. Hypnotized, I closed my eyes and cradled my head on a bench. I understand how Christ could sleep through a storm. Franco shook my shoulder.

"If you have come to sleep you should have stayed ashore," he said. The men wanted me to stay awake with them. Franco was feeling ill at ease, "because we took few tuna and because there is a lot of work ahead. You push me, I push you, to get the work done."

But I was so sleepy I couldn't keep my eyes open. "The sea lulls me into unconsciousness," I said, and lay down again. Franco covered me with a poncho.

I could not stay long enough to see the trap dismantled, but I would see the fourth and final mattanza of 1994. A giant white bluefin still swam in the trap. The sun was hot, there was no wind. Stefano, the barman, came with his father from Naples, who offered me biscotti and a thimble of hot espresso while we were pulled slowly out to sea.

The rais had invited his second son, Gaspare, and his teenage friends aboard the musciara. Gaspare wore a gold chain with the largest great white shark tooth pendant I'd ever seen. The men in the musciara removed their caps; Rosario shouted the litany and asked Saint Peter again for a good catch. "May God make it so!" the crew shouted in one voice.

The sun laid a wavy ribbon across the swells. Then a fog bank descended on us and obscured Sicily. Everything disappeared but the boats, the nets, the fishermen, and the spectators. The mist bound us together. It softened our voices and focused us on the square of death between us. By 10:00 A.M. all the fish had passed into the Camera della Morte. Forty men, five teams of eight, bent to pull up Il Coppo, and they began to sing.

But at the first verse four helicopters appeared. Like apocalyptic beasts, first they hovered, then they circled the Chamber of Death. Cameramen leaned out the doors of two of them. The noise was deafening, that hateful whumping; the propellers were drowning out the song.

The woman next to me joined me in giving them the Sicilian version of the finger: the Arm. Everyone on the vascello yelled and waved until the helicopters peeled off.

The dirge filtered through the fog again: "A promise is a debt! This God must help us and send us salvation, calm sea and wind at our backs, to find a safe port, to let us find shelter."

When the last tuna was taken Gioacchino walked across the taut net up to his knees in the sea to collect the tiny *vope* caught in Il Coppo's tight weave. These he stuffed into his wet pockets to feed to the feral cats.

By one-thirty the men were back at work in the Camparia. Antonio Casablanca and Giuseppe Aiello cut up bull fat to boil for greasing the boat slides. The dark blue boats would soon be dry-docked and white yachts would take their berths. I was leaving the next day. Antonio Casablanca said the salpata would start in earnest the next morning. I could go with them and be back in time to catch the afternoon ferry.

Weigh anchors and leave, the last two things on my list.

In the Camparia's courtyard, Gioacchino asked, "Are you really leaving tomorrow?"

"Yes."

"Wait here," he said, and went off on his bike. He came back with a purple velvet bag. He held it in one hand and dumped its contents into the other. Out spilled a thick gold chain weighted by a monstrous white shark's tooth framed by a golden heart. It was twice the size of Gaspare's. Gioacchino slipped the chain over his head and let the ivory tooth dangle on his bare chest. He tucked in his chin and struck a pose. I took his picture: Gioacchino, tooth and heart.

I packed that night, and next morning I was at the dock early; some men were already unloading a part of the Costa Alta that had been damaged by the tanker. The net rasped across the asphalt, crawling back from the sea, gasping as the men pulled it ashore. It steamed like wet seaweed in the sunshine. They sang snatches of song, Clemente and Ferdinando trading impromptu verses that poked fun at their colleagues. I noticed the rais's musciara still tied up at the quay and wondered who was counting the fish.

"Once the salpata begins the rais no longer checks the trap," Clemente said. The men first dismantle the barrier nets and leave the trap until last. If there are fish in it when the last moment of the season has come, the rais can order a last mattanza, *una mattanza di salpata*, or he can let them go.

I had a talk with Clemente.

Un'avventura is the Italians' word for a brief romance, an adventure. That's what Clemente called it after ours was over.

"Why did you lie to me? You said you had no feelings for Margo. What happened?"

"I'm in love with her again."

"What changed?"

"It was the distance."

Absence made his heart grow fonder. I let him go—not altogether gracefully—and we're still friends. Clemente is one of a kind.

That last morning I was aboard the *Portopalo*, captained by Gioacchino, far off the northern coast of Favignana. The ship had broken Orlando's Sword. Gioacchino pointed to a marker. "The last floats are here," he said, and set to work on the twisted mess.

That year they'd taken 763 tuna, 133 of them during the fourth mattanza. In addition, throughout the season 77 tuna had suffocated in the trap. The tonnaroti watched themselves fading away with the bluefin.

I locked a copy of Fadwah Tuqan's poem in the wall vault behind a picture in my dining room for the next tenant to find. I handed the house keys back to my landlord, who gave me a lift back to town. Stefano stood me a drink, and Clemente put my suitcase in the backseat of his tiny Fiat and drove me to the quay. I waved from the top deck; he waved and drove off. While the island was still dreaming its spring dream, I went back to another world in a small green valley, far from any sea.

25

Interlude

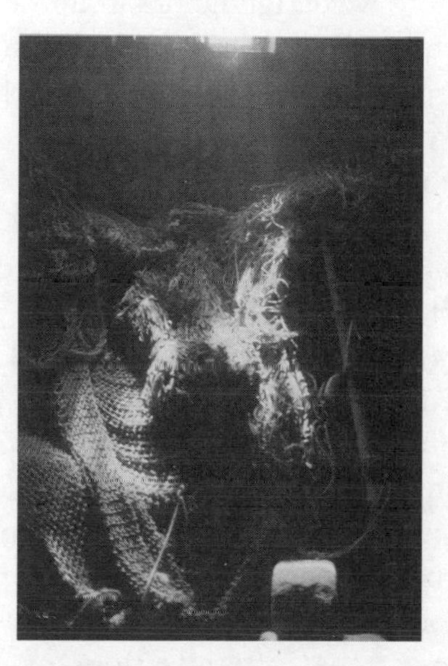

Favignana's bare pink rock receded from consciousness like a half-remembered dream. Vermont is green, with wooded hills and lush pastures. In summer clear streams flow around the knobby roots of trees clad in moss soft as moleskin. Mornings, the dew beads on the grass. In the fall mushrooms grow in all shapes and colors. Squat red toadstools glow like coals in the hollow of a rotten stump. A patch of yellow-stemmed jelly clubs with hunter green caps glistens in the mud along a brook.

The maples blaze yellow, red, and orange, prelude to the death of summer. Vermont's roads clog with "leaf peepers" who come to see the trees flare before they sputter out.

In winter my friend Jackie and I ski through her woods over snow fleas and pine needles sprinkled underfoot. In our imaginations we design fabrics from the textures and colors of nature: silk scarves with the shimmery iridescence of European maple trunks; wallpaper from a peel of pink birch bark; cashmere dyed the velvety brown and gray of shelf mushrooms clinging to the shady side of downed trees.

That January, when it was so cold I could not remember spring, I bought a ticket to Favignana. Ten days on my island. Favignana in winter was as sad as Occhiuzzi once told me it would be: "Four cats in the piazza, a couple of beer cans rolling around." No tuna, no tourists, and it rained a lot.

I rented a small apartment in Via Mazzini across from Zu Isidoro's bicycle shop. Night fell in late afternoon. From my second-floor window I looked down on the wet paving stones at the soft reflections of wrought-iron street lamps.

In the winter the Favignanesi stay indoors, play cards, and cook at home. Gioacchino plays basketball with the prison inmates once a week and hands out autographed color postcards of himself and Clemente killing tuna.

On clear days the sun warmed the stones of the Camparia's courtyard wall, its great wooden doors closed and locked behind the iron grille. But one day I found them open. The rais was alone in the gloom of the net house, making a new net from scratch. He sealed the frayed edges of the nylon cords over the flame of a single white candle. Already the town's thoughts slouched toward tuna season. Outside in the light, pearl gray gulls lifted from the boathouse roof and pressed against the wind. The mountain that I'd left pink and gray last summer was now sprouting green fleece. The parched hayfields that were harvested in May were green in January. And on the south side of the island, under the castle, a field of white and purple wildflowers grew lush and knee-high.

I forgot that the Two Columns was closed Tuesdays in winter, and when I arrived I found the door locked and the chairs stacked. I felt shut out. I had my coffee alone at the Bar del Marinaru at the port and watched the ferry unload. When the last passenger debarked the show was over. I biked to the fish market and watched a squid die.

Its gelatinous body rippled with shifting iridescent hues, a skin of pixels that pulsated in rolling waves of purple and periwinkle, chevrons of maroon and grass green, tiered rings of violet and red that flowed down its body slow as honey. The squid blazed through this desperate repertoire, all the patterns and colors it had left. Spent, it turned indigo, then black.

One day I found Rino, Clemente, and Angelo standing around a pot of boiling water on a gas burner in the back room at the Two Columns. The doors were closed against the cold, the windows fogged up. Clemente was about to boil an octopus for breakfast. Rino got knives and napkins and cut up a few lemons whose fragrance filled the room.

"This is the beauty of living on Favignana," Clemente said. "Write that down."

One night the following spring I saw a man beat up my girlfriend. Katria was another single woman who lived alone in a remote spot on Favignana. I could see her house in the distance from mine. I had met her at the Two Columns at the end of the 1994 season, and we'd become friends right away. We had lifted our glasses. "Fifty-five days in the Camparia, and I came out alive," I said.

"Fifty-five days in the Camparia, and you came out *reborn*," she said. Katria was a woman of means yet she was humble and generous. She had learned to cook from her grandmother and would often show up at my door with a bag of groceries and an impromptu party. She was thirty-something and still single but had broken with tradition and left her parents and siblings on the mainland to live here alone in a house her father bought her. She'd been here for many years, and loved the island more than I did.

We swam, took the sun, pried limpets off rocks with pointy knives, and ate them from the shell. We would spend afternoons reading quietly in secret coves she knew. She never told me her boyfriend abused her.

Mommo was married. He'd been a tonnaroto; now he worked off the books in construction and sometimes dug graves, but mostly he frequented Katria's. He told his wife they were just friends. We'd eat and drink, at my house or hers; she'd bring her Neapolitan tapes and we'd sing along. One night when she was supposed to cook dinner for Mommo and his friends, she and I were out drinking limoncellos instead. And we were having too much fun to stop. When Katria ran out of money she drove back to her house to pick up more. I stayed in the car.

Mommo was waiting in the house. I heard yelling and things crashing. I ran inside. He was beating her above the hairline so the bruises wouldn't show; he knew her father was coming the next day to visit. He beat her so hard the gold ring Katria had given him flew off his finger and across the room. She was being punished for too much independence.

I remembered the keys were in the ignition. While she fended off his fists I ran out, got behind the wheel, turned the key, and drove, drunk, to waken the carabinieri. It was past midnight. I stood at the gate and yelled, "He's killing her! Come quick." A light came on in the second-story window, and two policemen leaned on the windowsill. "Mommo and Katria! He's killing her!" In a flash they were out, two handsome guys dressed in their civvies. This happened regularly, they said. One of them commandeered the car and drove us back to Katria's house.

The two cops pulled him off her, but as they held him Mommo struck out with his foot and kicked me in the arm. He said he'd kill me. "I'll stuff your body in the sewer!"

The next day I filed charges. The carabinieri were happy. "You're very brave," said the one who wrote down my account of the evening. He showed me the file they'd amassed on Mommo, an inch and a half thick. None of the charges stuck, though, because the witnesses always dried up.

Mommo didn't ask me himself to rescind the charges. Instead he sent Katria. She came to me pleading, tears in her eyes.

"Teresa, change your mind, please."

"You're better off without him," I said.

"If you press charges my father will find out and he'll make me leave Favignana. Would you take that away from me?"

In the end I couldn't. Favignana had been Katria's island long before it was mine; it was not my place to decide her fate. So Favignana had become the cause of her perdition. Katria stayed with Mommo, and I lost a friend.

26

Resurrection

The next spring the ciurma was sullen and depressed. Again there was talk that Castiglione might not renew his contract for the next year. Who would run it then?

The Queen of Tonnaras had passed from the hands of the Arabs to the Norman kings, then to nobles, to billionaire merchants, and now to an absentee owner who leased her to a local businessman who might not want her anymore. Who could take the tonnara? Who would save it? It had survived twelve centuries; there had always been someone with the capital to trap bluefin and make more money. But now the fish were few and small, and the profits were dwindling with them. I left Favignana with a sense of foreboding.

ꝏ

The following winter I telephoned Gioacchino. It was January, and I looked through the glazed branches of a three-hundred-year-old maple down at the frozen mud while I waited for him to answer. I pictured him on his sunny island, his huge hand dwarfing the receiver.

Castiglione had not renewed his contract, Gioacchino said. At Gioacchino's urging, the tonnaroti had formed a cooperative and called it La Mattanza. They made a pact with the owner, Luigi Parodi, to run the tonnara and split the profits with him. Parodi would lease them the equipment and buildings for free. A few tonnaroti had risked their savings as collateral for a bank loan. And for the first time a ciurma had chosen its own rais.

Out of respect, the cooperative first offered the job to Salvatore Spataro, Gioacchino said, but he had declined it and gone to work for Castiglione at his tonnara at Bonagia, where they do not sing. Then the men elected Gioacchino their rais, and the president of their cooperative, and, apparently, their savior.

"I am rais on land as well as at sea," Gioacchino said.

My head spun with the changes. Then it hit me.

"Gioacchino, you are the eighth rais of this century," I said. I could picture his giant form standing in the musciara, surrounded by columns of saltwater. And then I saw his tombstone, and the four letters carved above his name, and I think he saw it too.

He would set the queen on her feet, but how long could she stand?

All through the late winter I watched the unexpected Comet Hale-Bopp do a slow blaze across the New England sky. When I returned to Favignana in April its sparkling wake had clouded and dimmed. The comet was streaking away, heedless, not to return for a few thousand years. The ferry *Simone Martini* passed the tonnara boats tied up in midsea for the afternoon check. "There will be a tonnara as long as there are fish," Clemente had said.

I walked to the Camparia and stowed my bag outside the rais's office. No one was there. I went to Rosetta's house for lunch—spaghetti with hot pepperoncini and fava beans—then back to the Camparia. Capoguardia Messina was at the gate. His chunky four-year-old-grandson was clinging to his pants leg; the boy followed his important *nonno* around with obvious awe and affection. Messina spread his hand on the child's head.

"This is my last season," he said. The tonnara would lose the expertise of the oldest working tonnaroto. "After this year I go out."

Aiello, the gatekeeper, had already retired, but he left a shiny memento under the old kitchen chair where he had kept watch. I bent to pick up the two-hundred-lire coin but found it cemented to the ground. Somebody laughed. It was old Capiletto, a head shorter than me, blue eyes sparkling in two crinkled slits. He linked his arm in mine and strolled around to show me the changes with obvious pride.

Castiglione had furnished his leased tonnara well but had taken everything with him when he left. The dark alcove between courtyards was filled with hundreds of new plastic buoys in red, orange, and yellow, bought with the cooperative's borrowed money to replace Castiglione's. Girolamo was still diving for the ciurma, but now he had to use his own scuba gear. The ziggurat of stone anchors, eight thousand creamy white blocks, towered twice as high as last year; Castiglione had taken his miles of chain with him, and now they needed the stones for weights. These were native tufa from Favignana's last quarry. Capiletto slapped a block. "These won't melt," he said.

Coils of steel at the mole and many anchors still on the beach told me the tonnaroti had not finished setting the barrier nets.

The fleet was black again. The freshly pitched boats in the harbor were the Florios' original vessels, wooden dinosaurs dry-docked these eleven years. Parodi had sold these old boats along with the cannery to the Sicilian region as part of a planned tonnara museum. But Gioacchino had convinced the region to lease him the boats for three months, which it did, at no cost. Gioacchino had turned the museum pieces into a working fleet. Their lugubrious sheen reflected the ghostly threads of light in the water. The black boats nodded heavily in port, grave and patient.

In the gatekeeper's chair sat the old Rais Gioacchino Ernandes, who hadn't set foot at the Camparia for eleven years, the entire tenure of his successor. Seeing me, he wagged his hand. "Come here," he said, imperially, I thought, and welcomed me back.

"Have you seen what that man did to the tonnara?" He meant Salvatore Spataro.

"What has he done?"

"All the boats, rotten. All the equipment, gone. The old nets in a mess, the ropes rotten."

"Was it his duty, or the duty of Castiglione, to keep all this in good shape?"

"He was the rais, it was his responsibility."

"But why didn't he do it? Was it out of vendetta, or was he just not suited for the job?"

"He was not suited, no. But he did it for vendetta."

"Vendetta against whom? He is also a Favignanese."

"Against Parodi. He took sides with Castiglione. They worked together. Castiglione wanted to ruin Parodi and take over, to buy the tonnara of Favignana. Now Spataro is Castiglione's slave," he said, an ordinary tonnaroto in another man's ciurma. It was hard to imagine his abdication.

"He comes and goes. He doesn't pass by here. A slave," Ernandes said, quite complacent.

Then Bertuzzo appeared, the man in his eighties who had sung me old songs in a reedy voice with a World War II guitar at the pensioners' shed. Bertuzzo had been one who quit in solidarity with Ernandes when Castiglione had ordered the trap moved. Other ciurma alumni, men who had kept their distance for more than a decade, began to collect at the gate, timidly feeling their way back to the Camparia, their alma mater.

One looked around the courtyard and said to no one in particular that he'd "worked fifty years inside here, under four or five raises," including Ernandes, his father, and Salvatore Mercurio. This year even the octogenarians had come in the tonnara's time of need to sew tears and join seams.

But I also saw many new faces, young bachelors without a wrinkle who had replaced the older family men who'd quit, unsure of the cooperative's ability to pay them. Most of the new boys didn't even know how to tie a knot; often they got in the way and exasperated the old hands, but they were necessary and eager, a new generation ready to learn this ancient trade.

The net house, the men busy in the yard, the anchors on the beach, the shiny black boats—everything seemed infused with new life welling up from deep roots. There had been a revolution. For the first time in history half the fruit of the tonnara now belonged to the fishermen. This was the source of their pride.

A gnarly hand held up before me three skeleton keys on a ring. "These are the keys to the Camparia," the old Rais Ernandes said, "to the palazzo, and to one other building belonging to the Parodis. Parodi told me, 'When I am not present, you are Parodi for me.'"

"How does the rais react to that?" I asked.

He didn't look at me, but over my shoulder out to sea, where he expected Gioacchino Cataldo to appear on the horizon.

"He wants me well. He says I am his father."

The new rais was out setting La Coda and might not be back until dark, he said. The rectangular trap was already in the water, set a little closer to land than it had been for the last eleven years. Gioacchino might not be able to complete the *campile* tonight, "because it is complicated and there are many new hands," Ernandes said.

In the morning I found Gioacchino in the rais's office. He had taped a mint-condition Florio tuna can label to his door frame and tucked a fresh palm crucifix behind the Sacred Heart at his back, but otherwise things looked the same. In the gloom behind the rough wooden desk he looked like a trapped bear.

He had used two of his houses as collateral for a bank loan and had convinced others to risk their savings to keep the tonnara afloat. Somebody said, "There ought to be a statue to him, in bronze. Because if it weren't for Gioacchino, it would be all over."

But others now wanted assurance that they would be paid. They feared, perhaps with reason, that they wouldn't see a paycheck until the season was over and the expenses were paid, and there might be nothing left for them that year. They were living day to day as it was and couldn't afford such a long-term gamble. Many were ready to bolt; some already had. The tonnara was tradition, yes, but they killed tuna to stay alive.

A clutch of doubters had formed outside the office door. Gioacchino emerged and gathered the men around him. He explained that he was trying to finalize the cooperative's contract with Parodi. Gioacchino was in the new and irreconcilable position of being a rais "both on land and at sea." He was their boss and their union leader. The world was changing again. He spoke softly but in a tone of command, torn between a desire to speak as a cooperative founder and a longing for the authority of the raises of old. They would have brooked no complaints.

The tonnaroti worked at sea until nearly ten that night, setting the rest of La Coda. But next morning I saw two of them lingering at the Two Columns long after the eight o'clock break.

"What's up?" I asked.

"The sea is rough today."

I had been to the port, and the sea was calm.

"You mean far out to sea?"

"Also on land," the tonnaroto said. The cooperative still had no contract with Parodi. The men were staging a job action. At the Camparia the ciurma dawdled and loafed, truculent, waiting for Parodi's representative to arrive from Tràpani with a contract.

Gioacchino called the men into the courtyard to plead for their faith and cooperation. The nets had to be put in the water soon if they were to trap fish. He reminded them that last year Castiglione hadn't signed a contract until July. Rocco, who had campaigned with Gioacchino for a cooperative, made a speech from the back of the crowd.

"I will work with or without a contract, even without pay, because the tonnara is in my blood," he said. He held up his arm and pointed to a vein. Gioacchino left it at that and disappeared for a while. When he came back the men were working again on the nets, joking and calling to one another. "Acqua! Ferma lì! Molla!"

The Coppo they were now preparing hadn't seen water for twenty years. They had salvaged it from the abandoned Formica tonnara. It was of the same dimensions as the old Favignana Coppo, but heavier, with a mesh so tight it held water. They unfurled it in sections as old Rais Ernandes and his former lieutenant paced it shoulder to shoulder, hands clasped behind their backs, weighing it with their eyes. Other old-timers did likewise.

"This is *ponente*," the western end, said Pasquale Messina, a pensioner who had served under Ernandes. His fellow alumni, lined up like crows along the low stone wall under the palm tree next to the gas station, were watching. They were the old guard in their pilly go-to-church cardigans and loafers and clean white socks.

Under the eave of the boathouse Vito cut lengths of nylon rope with an acetylene torch while baby swallows cheeped in their nests above his head. "Love doesn't die," he crooned. The tonnaroti sat on Il Coppo cross-legged, humped over, sewing seams. It was peaceful work and meditative. The men called to each other over the undulating expanse of net. A dimpled cloud of organdy dimmed the sun. Tomorrow they would raise the saints, and Gioacchino would preside.

Gioacchino lashed red lilies and white calla lilies to the cross, which now lay in his musciara. He had added two new faces to the saints: Padre Pio and Pope John XXIII, both uncanonized peasants revered as saints just the same for their kindness and humble roots. Above them all was Saint Ann, Jesus' grandmother, the patron of women in labor and the wife of Joachim, whose namesake now husbanded the tonnara.

The sea near the trap entrance was soft and silken. I was in a little blue boat, Gioacchino's own, which he had loaned to the fleet, with Alberto rowing slowly. The water dripping from his oars was like birdsong. Gioacchino raised the cross and held it as the men secured it. He recited the words of the litany for the first time as rais. I read his sweatshirt, printed in English, and wondered if he knew what the words meant: "We finally made it."

When the time came Gioacchino asked the old rais to ring the bell, then he signaled for a blast on an air horn so they could hear it in the piazza: tomorrow mattanza. As a horde of journalists pressed around him, Gioacchino was interviewed on national television and the tonnaroti cheered in the background.

"We have to be a family," the rais said. He had his arm around Ernandes, who stood poker-faced beside him. "What we did has never been done in such a short time. Everything is in the best working order. I wish to everybody that all goes well, and that we capture fifteen hundred tuna."

He paused after the sound bite, then said, "The tonnara belongs to Sicily, to Italy. It is a national patrimony. Tomorrow the tourists begin to arrive to see the mattanza. We will have eight to twelve mattanzas. We augment tourism by 15 to 20 percent." In an aside to the press he promised to keep the television cameras close to him during the mattanza.

"We are content," Gioacchino said to the cameras. "I do this with care and a serene soul. I am an honest person. I could live without the tonnara quite well. I do it for this island. I am Cataldo, the new rais. Come to Favignana."

Gioacchino gave the tonnaroti the dress code for his first mattanza: no Rambo sweatbands, no T-shirts with advertising on them. "We will have an old-fashioned mattanza," he said. Rais Ernandes invited the ciurma for ice cream at a place around the corner, sixty-three cones, his treat.

The next morning Rais Ernandes was rowed in state around the Chamber of Death with Rais Cataldo in the musciara, but the senior rais debarked just before the mattanza, leaving his heir to direct the killing alone. They took 113 fish in Gioacchino's first mattanza.

When it was over the tuna were not towed to Tràpani. Instead Gioacchino sent the bluefin to bleed on Favignana. Red puddles stained the stones at the mole where a crane lifted the tuna three at a time from the vascello. It gripped them by their lunate tails; they hung like grape clusters. A crowd gathered and saw that their tuna were silver and blue.

Men in satiny jackets emblazoned "Eurofish" loaded them into refrigerator trucks that would cross the strait on the ferry, then take them to Marsala. Capoguardia Messina stood next to the truck, tying knots in a rope as the tuna were loaded, and Leonardo wrote the numbers down in a book. Before they left, U Straviatu, the fishmonger, managed to buy a 176-pound bluefin. Later a line formed in the street outside his shop.

The rais had changed into clean green pants and sneakers and biked up to inspect his harvest. This was his triumph. He had snatched the tonnara from the fire and just commanded his first mattanza. He had recovered what had been lost; Gioacchino brought the life-giving elixir back to the island.

His wife arrived in a pretty chiffon dress. The crowd pressed closer. Shy before the people, she congratulated the rais by shaking his hand. Then she smiled up at him, and Gioacchino bent to kiss her.

Survival

Once, the tuna snares thrived in Algeria, Corsica, Tunisia, Malta, Dalmatia, and Turkey. In Portugal they were called *armaçoes*; in Spain, *almandrabas*; in France, *madragues*. The cause of abandon: insufficient fish to make a profit.

Once, there were tonnaras all over Sicily. I can chant their names:

White Fountains; River of Noto; Be in Peace;
Vindicari; Capo Negro; Guzzo; Capo Bojuto;
Magazzinazzi; La Sicciara; Ursa; Carini; Capace;
Isola delle Femmine; Mondello; LoMonaco;
San Giorgio de' Genovesi; Tonnarazza; Acqua Santa;
Capicello; Felice; Salsa; Punta Secca;
Punta Rais; Scopello; Monte Rosso;
Gaffi; Puzallo; San Nicola-Malastri;
Camanara; Tre Fontane; Tono di Sciacca;
Mazzarelli; Porto Palo; Cofano; San Vittore;
Nubbia; La Monzella; La Boeo; Cannizzo;
Santa Maria la Nuova; Mazzara; Capo Banco;
Acireale; San Calogero; Poggio Grosso;
Brucoli; Mililli; Magnisi; Costa de'Tuoni;
Sant'Elia; Solanto; San Nicolo l'Arena;
Castellamare del Golfo; Vergine Maria; Arenella;
Formica; Trabucadu; Saline;
Pittinuri; Flumentorgiu.

Gone, all gone.

By 1998 Favignana was my second home. I knew more people there by name than I did in my own hometown. When I visited I stayed with Rosetta. I boarded the hydrofoil at Tràpani. A white haze hung over the water. The sun was hot, the sky streaked with low white clouds. I had three days on Favignana, and I hoped to see a mattanza.

A month-old news clipping from *La Sicilia* was tacked to the Camparia wall inside the courtyard. Thirty percent of the ciurma is new, it said, the unemployed youth of Favignana. So even more of the old hands had left.

The cooperative was still in debt despite the receipt of seventy-five million lire in subsidies from the federal government, forty-five million lire from the province of Tràpani, and thirty million from the commune of Favignana, "and we still wait for that of the region," the rais said in the article.

Gioacchino betrayed no romantic illusions about the survival of the tonnara. "The intention was that of transforming the mattanza into a tourist event, and we succeeded," Gioacchino was quoted as saying.

Rocco was a boat captain now. At one-thirty I boarded his vessel to check the trap. The towboat pulled us out to sea, and Rocco's crew rowed us to I Santi. The old black boat was battered and sticky with warm pitch. Rocco handed me the spyglass. I leaned over the gunwale and pressed it into the sea, letting it bob with the waves. The electric blue life stew slipped through the net's diamond weave. A purple jellyfish throbbed in my face. A small tuna passed through a shaft of sunlight.

"You watch, Teresa. Eh? Because I'm going to sleep a little." Rocco curled up in his boat. "If there are tuna, you tell me, okay?"

I did see one more mattanza that year. A little before seven I stood before the Camparia's gate, apart from the small crowd that had gathered. The tourists clumped nervously beside the long black vascello, waiting for permission to board. Today some would see their first mattanza. They came for a thrill, to see something as huge and wild as the bluefin meet its death. They came to see something their friends could not quite explain to them.

They came not knowing they would witness an ancient rite that connected them to their own past, and to their common destiny. They might look at the bluefin and see themselves swimming again where they were born, struggling, begetting, and dying in the same room.

I stood alone, surrounded by beauty. The bay before the boathouse was impossibly green. The sea beyond was a deep blue field run through by mysterious currents of a lighter color, marine highways that have formed and disintegrated every spring. Santa Caterina, the mountain and castle, was in a simple yoga pose, standing, gripping the ground, gazing ahead, sending breath to the body, casting out all thoughts. To my left the cannery wall concealed all inside but the smokestacks that rose above it. Behind me stood the white sandstone wall of the net house, its alcoves now empty under the vaulted ceilings, its brick floor polished to a soft patina by the shuffle of many feet. Below it, the roomful of rainwater.

The bluefin had passed before this island before the Camparia was built, before Ruggero made his castle, before Carthage plied these waters, before Favignana had a name. They had to come. They come still, seduced by the light and by the water, its warmth, its clarity, its salt. Their ranks have thinned since they passed Gibraltar. They have been tracked and encircled by purse seine nets; the oldest and the strongest of them have been taken. These few swim into the web of unseen threads. They mill and mate inside its walls, clouding the water with their roe and milt, oblivious to their fate.

The trap was in the water; the fish were in its sixth room, and the people were going to it.

We boarded and became part of the ancient, somber procession. Marino shouted the prayer into the wind. When he finished the men in the musciara put their hats back on. The sky was white but for one strange cloud that settled like a dark hen over the castle and hid it.

Vincenzo Sercia's own son, Nino, a new tonnaroto, was the new First Voice. He knew the songs by heart. The songs made them worthy to kill the bluefin. He sang; the melody was a plaint, a rolling wave of lament. The chorus answered: "Ai-a-mola, ai-a-mola." What does it mean? What door does it open? He sang of the birth-giving Virgin, of God who created the sun and moon, all the people, and the fish in the sea.

The fish were beautiful, just fifty or sixty of them, but most of them giants, with no small ones. The creatures in the water were symbols of the life of the unconscious, Campbell said, primal forces that had to be tamed. Seven hundred, eight hundred pounds each, they raged and seethed; the splashes they made went ten feet high. The teams of young tonnaroti, pressed into killing their first year, could not lift the largest giants. They would yell, "Uno, e *due!*" and heave, but the giant always slid back in the water.

Rocco had repainted his gaff orange and sharpened its hook to an icepick's cruelty. But one giant he could not lift. Twice he lost his grip on the gaff and dropped it in Il Coppo, where the giants were flailing; Angelo, Clemente's son, dived in to retrieve it. Another giant swam off with a gaff buried in its flesh. It caught under the rais's musciara, and again Angelo dived to free it. He waded on the net waist-deep in water, grabbed the fish, and brought it gasping back toward the vascello.

The crowd had already applauded, but this one was still alive. Rocco and Angelo, at the center of their killing team, could not position it upright at the side of the boat. To get a better grip, they slammed their gaffs into its side, but then the flesh would rip and they'd lose their grip and have to gaff it again.

The fish and Rais Gioacchino could endure this suffering only so long. At last the rais screamed at the tonnaroti. Angelo dived again, grabbed the tuna by its fins, and pulled it down to his father's team. Clemente and Benito lifted it on the first try.

Epilogue

I could not go to Favignana in the spring of 1999. By the time I called Gioacchino to find out what had happened, the maple at my window was turning yellow.

They took 1,100 tuna, he said. Their average weight was 110 pounds. The largest weighed 550 pounds, 286 pounds less than the largest of the year before.

"We started too late; the biggest had already passed," Gioacchino said. Next year, if there is a tonnara, the saints must be raised by the first of April.

"What has become of Salvatore Spataro?" I asked.

"He became the rais at Bonagia. They took fifteen fish." An unusually strong current had sunk one of Castiglione's boats, loaded with nets. No one was drowned, but Spataro lost the season and spent the rest of the spring weighing anchors.

"There is a God," Gioacchino said.

I counted my mattanzas on a stiff coconut cord I found on the ground in the Camparia courtyard and tied the string to my day pack. Fifteen knots, every one a day in spring on Favignana. One morning as I left the Camparia the cord caught in my bike spokes and snapped off near the top. The force yanked me backward; I was stopped in my tracks. All those years, no, all those moments with my face pressed into the sea, broken off with a single stretch.

I keep what remains—the roses pressed in books, the stone fish on a shelf, a chunk of sea pine on my desk. A palmful of coral pebbles, a poem, a thousand pictures. I have a shard of Phoenician glass. A fragment of a fisherman's heart. And the map of a trap where the bluefin swim.

Photo Captions

Author's Note: The killing.
Chapter 1. My first mattanza, June 2, 1986.
Chapter 2. Rais Salvatore Spataro caresses the sea.
Chapter 3. The Chamber of Death on the morning of a mattanza.
Chapter 4. Near Cala Rossa.
Chapter 5. Interior of the net house.
Chapter 6. "The Tuna Merchant," a Greek Sicilian vase, fourth century B.C. Reproduced by permission of Museo Mandralisca, Cefalù, Sicily.
Chapter 7. View from "la giungla" in the cannery where tuna were once beheaded and hung to bleed.
Chapter 8. Cannery interior with copper cooking cauldrons to the right.
Chapter 9. An anchor boat.
Chapter 10. Rosetta Messina's altar to Saint Joseph.
Chapter 11. Loading a net.
Chapter 12. Gravestone of Rais Gioacchino Ernandez.
Chapter 13. 1761 engraving by Antonio Bova of a Tràpani tuna trap.
Chapter 14. A bluefin drawn up in Il Coppo.
Chapter 15. A detail of the cross bearing the saints.

Chapter 16. Rais Salvatore Spataro peers through a spyglass at trapped fish.
Chapter 17. Gioacchino Cataldo, tooth and heart.
Chapter 18. A creation of Rosario Santamaria, "Zu Sarino."
Chapter 19. Vincenzo Sercia holding forth at the pensioners' shed.
Chapter 20. Rosa Manuguerra at the cemetery.
Chapter 21. Il Coppo loaded in a longboat.
Chapter 22. Pasquale Mercurio, a retired tonnaroto.
Chapter 23. La Mattanza. Photo by Antonio Noto.
Reproduced with permission.
Chapter 24. La salpata began early. The tonnaroti pull the twisted mess aboard.
Chapter 25. Il Coppo in its winter lair in the net house.
Chapter 26. Rais Gioacchino Cataldo presides at the raising of the Saints.
Chapter 27. Monte Santa Caterina towers over the cannery
Epilogue. Gulls follow the musciara.

Selected Bibliography

Ariolo, Angelo. *Le Isole Mirabili*. Torino: Einaudi, 1989.

Block, Barbara A., Heidi Dewar, Charles Farwell, and Eric D. Prince. "A New Satellite Technology for Tracking the Movements of Atlantic Bluefin Tuna." *Proceedings of the National Academy of Science USA* 95 (August 1998): 9384–9389.

Block, Barbara A., Heidi Dewar, Tom Williams, Eric D. Prince, Charles Farwell, and Doug Fudge. "Archival Tagging of Atlantic Bluefin Tuna (*Thunnus thynnus thunnus*)." *MTS Journal* 32, no. 31: 37–45.

Campbell, Joseph, and Bill Moyers. *The Power of Myth*. New York: Doubleday, 1988.

Cataliotto, Alessandro. *Favignana: Memorie, note, ed appunti con speciale riferimento al Castello di S. Giacomo*. Girgenti, 1924.

Consolo, Vincenzo. *La Pesca del tonno in Sicilia*. Palermo: Sellerio, 1986.

Craft, Lucille. "$20,000 for One Fish?" *International Wildlife* (November-December 1994): 18.

Dalby, Royce. "Where Fish Are King." *Ad Astra* (February 1991): 27.

Erickson, Paul. "Tale of a Tuna." *Animals* (July-August 1994): 16–17.

Dorfman, Andrea, Gavin Scott, Dick Thompson, et al. "Depletion of Fish in World's Seas." *Time*, April 4, 1994, 71.

Finley, M. I. *A History of Sicily: Ancient Sicily to the Arab Conquest*. New York: Viking, 1968.

Fodera, Vito. *La Madrague Sicilienne de course*. Rome: United Nations Food and Agriculture Organization, July 1961.

Garrell, Martin H. "The Pursuit of Bluefin Meets Reality." *Sea Frontiers* (November 1993): 18.

Gillett, Robert. "Traditional Tuna Fishing: A Study at Satawal, Central Caroline Islands." *Bishop Museum Bulletin in Anthropology I* (1987): 6, 29, 30.

Guarrasi, Giuseppe. *Le Cialome delle tonnare*. Favignana: self-published, 1975.

Hemingway, Ernest. "At Vigo, in Spain, Is Where You Catch the Silver and Blue Tuna, the King of All Fish." *The Toronto Star Weekly*, Feb. 18, 1922, p. 15.

Homer. *The Odyssey*. Translated by E. V. Rieu. New York: Penguin Books, 1946.

Jones, F. R. Harden. *Fish Migration*. New York: St. Martin's Press, 1968.

International Commission for the Conservation of Atlantic Tunas. *Report for Biennial Period 1994–95*, pt. 1, vol. 1, English version. ICCAT, Madrid, Spain.

Jung, Carl Gustav. *Man and His Symbols*. Garden City, N.Y.: Doubleday, 1964.

Liddell, Henry George, and Robert Scott. *A Greek-English Lexicon*. New York: Harper and Brothers, 1883.

MacKenzie, Debora. "Too Little Too Late to Save Atlantic Bluefin." *New Scientist* (November 20, 1993): 11.

McKeown, Brian. *Fish Migration*. Beaverton, Oreg.: Timber Press, 1984.

Mather, Frank J. III, John M. Mason, Jr., and Albert C. Jones. *Historical Document: Life History and Fisheries of Atlantic Bluefin Tuna*. Technical Memorandum NMFS-SEFSC-370. Miami, Fla.: National Oceanic and Atmospheric Administration, 1995.

Mazzarella, Salvatore. "I luoghi e la memoria." *Kalós Luoghi di Sicilia*, no. 17 (November-December 1994): 9.

Ministero di Agricoltura, Industria, e Commercio. *Atti della commissione reale per le tonnare*. Rome, 1889.

Parona, Corrado. *Il Tonno e la sua pesca*. Memoria 68. Venice: R. Comitato Talassografico Italiano, 1919.

Partridge, Brian L., Jonas Johansson, and John Kalish. "The Structure of Schools of Giant Bluefin in Cape Cod Bay." *Environmental Biology of Fishes* 9, nos. 3 and 4 (1983): 253–262.

Proctor, Paul. "Robo-tuna." *Aviation Week*, November 7, 1994, 17.

Racheli, Gin. *Egadi: Mare e vita*. Milan: Mursia, 1986.

Safina, Carl. "Bluefin Tuna in the West Atlantic: Negligent Management and the Making of an Endangered Species." *Conservation Biology* 7, no. 2 (June 1993): 229–233.

Sarà, Raimondo. *Sulla biologia dei tonni (*Thunnus Thynnus L.*): Modelli di migrazione ed osservazioni sui meccanismi di migrazione e di comportamento*. Boll. Pesca Piscic. Idrobiol. (1973).

———. *Tonni e tonnare*. Tràpani: Libera Università di Tràpani, date of publication unknown.

Seabrook, John. "Death of a Giant: Stalking the Disappearing Bluefin Tuna." *Harper's* (June 1994): 48–53.

Smith, Denis Mack. *A History of Sicily: Medieval Sicily 800–1713*. New York: Viking, 1968.

Sonu, Sunee C. "Japan's Tuna Market." NOA-TM-NMFS-SWR-026. Washington, D.C.: U.S. Department of Commerce, National Oceanic and Atmospheric Administration, National Marine Fisheries Service, Southwest Region, September 1991.

Williams, Ted. "The Last Bluefin Hunt." *Audubon* (July-August 1992): 14–20.

Whynott, Douglas. *Giant Bluefin*. New York: Farrar, Straus & Giroux, 1995.

Zanca, Renato. "Tonni e ricchezza." *Kalós Luoghi di Sicilia*, no. 17 (November-December 1994): 2.

Zinnanti, Mario. *Cenni storici delle Isole Egadi*. Tràpani, 1912.

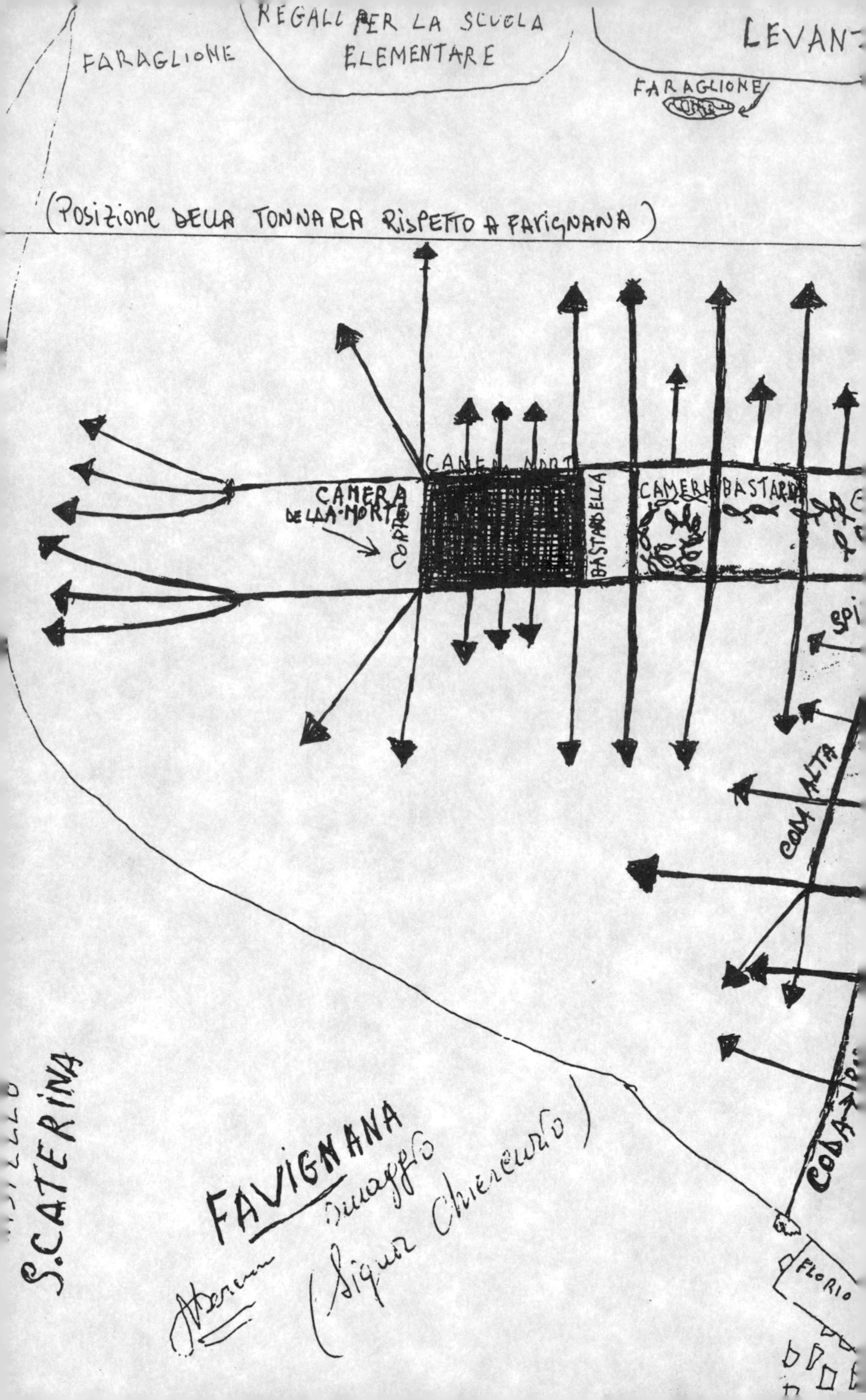

REGALI PER LA SCUOLA ELEMENTARE
FARAGLIONE
FARAGLIONE
(Posizione della tonnara rispetto a Favignana)
CAMERA DELLA MORTE
CORPO
BASTARDELLA
CAMERA BASTARDA
CODA ALTA
CODA
FLORIO
S.CATERINA
FAVIGNANA